AF361917

Gas Bearings: Modelling, Design and Applications

Gas Bearings: Modelling, Design and Applications

Editors

Terenziano Raparelli
Federico Colombo
Andrea Trivella
Luigi Lentini

MDPI • Basel • Beijing • Wuhan • Barcelona • Belgrade • Manchester • Tokyo • Cluj • Tianjin

Editors
Terenziano Raparelli
Politecnico di Torino
Italy

Federico Colombo
Politecnico di Torino
Italy

Andrea Trivella
Politecnico di Torino
Italy

Luigi Lentini
Politecnico di Torino
Italy

Editorial Office
MDPI
St. Alban-Anlage 66
4052 Basel, Switzerland

This is a reprint of articles from the Special Issue published online in the open access journal *Applied Sciences* (ISSN 2076-3417) (available at: https://www.mdpi.com/journal/applsci/special_issues/Gas_Bearings).

For citation purposes, cite each article independently as indicated on the article page online and as indicated below:

LastName, A.A.; LastName, B.B.; LastName, C.C. Article Title. *Journal Name* **Year**, *Volume Number*, Page Range.

ISBN 978-3-0365-5607-9 (Hbk)
ISBN 978-3-0365-5608-6 (PDF)

Contents

Editorial

Special Issue "Gas Bearings: Modelling, Design and Applications"

Federico Colombo *, Luigi Lentini, Terenziano Raparelli and Andrea Trivella

Politecnico di Torino, Department of Mechanical and Aerospace Engineering, Corso Duca degli Abruzzi 24, 10129 Turin, Italy
* Correspondence: federico.colombo@polito.it

Citation: Colombo, F.; Lentini, L.; Raparelli, T.; Trivella, A. Special Issue "Gas Bearings: Modelling, Design and Applications". *Appl. Sci.* **2022**, *12*, 9048. https://doi.org/10.3390/app12189048

Received: 8 September 2022
Accepted: 8 September 2022
Published: 8 September 2022

Publisher's Note: MDPI stays neutral with regard to jurisdictional claims in published maps and institutional affiliations.

1. Introduction

Gas bearings are widely employed in high-precision devices and in high-speed applications, such as in micro turbomachinery and micro machining tools. In metrology and high-tech industry, the absence of contact between the surfaces that are in relative motion is appreciated due to the possibility of obtaining high-precision motion. Recent trends in small-scaling turbomachinery demand increasingly higher rotational speeds; the advantage is the ability to operate with higher power to weight ratios. Additionally, in this case, gas bearings represent the only solution due to their simplicity and long operating life, with no need for maintenance.

This issue collects research and review papers covering the modelling, design, and application of gas bearings. Each published paper distinguishes itself from the others in the sense that it deals with the problem from an original point of view. Some papers analyze several prototypes to improve their performance, while others concentrate more on the mathematical modelling of bearings, developing analytical or numerical formulations with lumped or distributed parameters. These different approaches show how complex the analysis of gas bearings can be, but they also show how fascinating the process of finding simple solutions to complex problems is.

2. Outline of the Issue

The papers published in the Special Issue are here summarized in the order of their publication dates. Most of them investigate gas bearings using theoretical and experimental approaches. Numerical models, after they have been experimentally validated, are important tools for the design and optimization of gas bearings. Current models are able to consider multi-physical problems such as fluid–structure interaction, thermal effects, and porous materials. In some cases, tailor-made software is written to solve the equations, and in other cases, commercial software is employed. The experimental results confirm the validity and accuracy of multi-physical numerical models.

In high-precision devices, such as fly cutting machine tools or rotary tables, the deformation of solid parts is often not negligible, as it influences the performance of the aerostatic bearings. In [1], the fluid–structure interaction effect is considered to investigate the static and dynamic performance of an aerostatic spindle. In particular, the thickness of the thrust plate is optimized to reach a compromise between the reduction of the spindle mass and the increase in the thrust bearing stiffness.

Foil gas bearings are particularly useful in turbomachinery, as they are able to sustain high-speed blowers, compressors, and turbochargers without the requirement of an external air supply. Paper [2] describes the thermo-hydrodynamic model of a bump foil thrust gas bearing. The cooling models of the foil structure and thrust plate as well as the influence of temperature on the thermal expansion of the bearing structure are considered. The effects of the bearing speed, thrust load, and external cooling gas on the bearing temperature field are calculated and analyzed.

Paper [3] reviews the state-of-the-art of gas foil bearings in China, many of which have been developed for use in high-speed turbo machinery. Different types of foil bearings are described (bump, multi-leaf, wire, viscoelastic, spring and protuberant), as are their different applications (e.g., fuel cell air compressors, blowers, turbo-expanders, and oil-free turbochargers). The addressed challenges concern thermal management, rotor-dynamic stability, and wear-resistant coatings.

High-tech industry products such as solar panels, chips, and displays are composed of very thin layers of brittle materials that need to be handled with care. An original solution to handle the thin substrates without mechanical contact is proposed in [4]. The system works by floating the substrate on a thin film of air and by creating a viscous traction force on it. The actuator is improved to reach a bandwidth of 300 Hz, while the position control loop of the substrate reaches a bandwidth of 10 Hz with a position error of less than 13 μm. It is a good example of the so-called tribotronics discipline.

It was shown in the literature that the radial error motion of an aerostatic journal bearing is mainly influenced by the shape errors of the shaft. The influence of the shaft roundness and the cylindricity errors on the aerostatic spindle's rotation accuracy is evaluated in [5]. The research proposes an index that is more effective in predicting the spindle's rotation accuracy than the roundness and cylindricity.

Paper [6] presents an elegant analytical model of a porous gas journal bearing after a literature survey on the topic. The analytical model is able to predict the flow and the dynamic force coefficients. An FE model is compared to the analytical one, and a perfect match is found. A validation of the static results with experimental data from the literature is also carried out. The paper represents a useful design guide for the quick selection of physical parameters, both for static performance and for stability analysis.

A partial arc annular-thrust porous journal bearing that can carry both radial and axial loads is investigated in [7]. The analysis is carried out using commercial multi-physical software in order to evaluate the pressure distribution for low eccentricity as well as for low tilting angles and tilting speeds. The dynamic coefficients of tilting motion are then determined using the finite difference technique.

Finally, the modal parameters of an aerostatic spindle are identified experimentally based on the impulse response in paper [8]. Different methods are proposed to identify the damped natural frequencies and the damping ratios of the conical and cylindrical modes of the spindle. The same modal parameters are also evaluated with a non-linear numerical model, and a comparison with experimental results is provided to validate the model.

3. Future of Gas Bearings

The different gas bearing solutions require accurate mathematical models and cannot be easily standardized due to the complexity and variety of the problems. Despite gas lubrication being a well-established discipline since the 1970s, the ever-rising demands of modern technology offer new challenges to obtain higher precision and higher rotational speeds, stimulating researchers to broaden the frontiers of gas lubrication.

Author Contributions: Writing—original draft preparation, F.C.; Writing—review and editing F.C., L.L., T.R. and A.T. All authors have read and agreed to the published version of the manuscript.

Funding: This research received no external funding.

Acknowledgments: The editors would like to acknowledge all of the authors for their high-quality contributions to this Special Issue. The MDPI management and staff are also acknowledged for their continuous support.

Conflicts of Interest: The authors declare no conflict of interest.

References

1. Gao, Q.; Gao, S.; Lu, L.; Zhu, M.; Zhang, F. A Two-Round Optimization Design Method for Aerostatic Spindles Considering the Fluid–Structure Interaction Effect. *Appl. Sci.* **2021**, *11*, 3017. [CrossRef]
2. Liu, X.; Li, C.; Du, J.; Nan, G. Thermal Characteristics Study of the Bump Foil Thrust Gas Bearing. *Appl. Sci.* **2021**, *11*, 4311. [CrossRef]
3. Hou, Y.; Zhao, Q.; Guo, Y.; Ren, X.; Lai, T.; Chen, S. Application of Gas Foil Bearings in China. *Appl. Sci.* **2021**, *11*, 6210. [CrossRef]
4. Hooijschuur, R.; Saikumar, N.; HosseinNia, S.H.; van Ostayen, R.A.J. Air-Based Contactless Wafer Precision Positioning System. *Appl. Sci.* **2021**, *11*, 7588. [CrossRef]
5. Zhang, G.; Zheng, J.; Yu, H.; Zhao, R.; Shi, W.; Wang, J. Rotation Accuracy Analysis of Aerostatic Spindle Considering Shaft's Roundness and Cylindricity. *Appl. Sci.* **2021**, *11*, 7912. [CrossRef]
6. Andrés, L.S.; Yang, J.; Devitt, A. Porous Gas Journal Bearings: An Exact Solution Revisited and Force Coefficients for Stable Rotordynamic Performance. *Appl. Sci.* **2021**, *11*, 7949. [CrossRef]
7. Hwang, P.; Khan, P.; Kang, S.-W. Parameter Sensitivity Analysis on Dynamic Coefficients of Partial Arc Annular-Thrust Aerostatic Porous Journal Bearings. *Appl. Sci.* **2021**, *11*, 10791. [CrossRef]
8. Colombo, F.; Lentini, L.; Trivella, A.; Raparelli, T.; Viktorov, V. Experimental and Numerical Dynamic Identification of an Aerostatic Electro-Spindle. *Appl. Sci.* **2021**, *11*, 11462. [CrossRef]

Article

A Two-Round Optimization Design Method for Aerostatic Spindles Considering the Fluid–Structure Interaction Effect

Qiang Gao [1,2], Siyu Gao [1,*], Lihua Lu [1,*], Min Zhu [3] and Feihu Zhang [1]

[1] School of Mechatronics Engineering, Harbin Institute of Technology, Harbin 518057, China; gaoq@hit.edu.cn (Q.G.); zhangfh@hit.edu.cn (F.Z.)
[2] School of Astronautics, Harbin Institute of Technology, Harbin 518057, China
[3] School of Electrical Engineering and Automation, Harbin Institute of Technology, Harbin 518057, China; zhumin@hit.edu.cn
[*] Correspondence: gaosiyu@hit.edu.cn (S.G.); lihual@hit.edu.cn (L.L.)

Abstract: The fluid–structure interaction (FSI) effect has a significant impact on the static and dynamic performance of aerostatic spindles, which should be fully considered when developing a new product. To enhance the overall performance of aerostatic spindles, a two-round optimization design method for aerostatic spindles considering the FSI effect is proposed in this article. An aerostatic spindle is optimized to elaborate the design procedure of the proposed method. In the first-round design, the geometrical parameters of the aerostatic bearing were optimized to improve its stiffness. Then, the key structural dimension of the aerostatic spindle is optimized in the second-round design to improve the natural frequency of the spindle. Finally, optimal design parameters are acquired and experimentally verified. This research guides the optimal design of aerostatic spindles considering the FSI effect.

Keywords: aerostatic spindle; fluid–structure interaction; restrictor geometrical parameters; structural dimension; optimal design

Citation: Gao, Q.; Gao, S.; Lu, L.; Zhu, M.; Zhang, F. A Two-Round Optimization Design Method for Aerostatic Spindles Considering the Fluid–Structure Interaction Effect. *Appl. Sci.* **2021**, *11*, 3017. https://doi.org/10.3390/app11073017

Academic Editor: Terenziano Raparelli

Received: 12 March 2021
Accepted: 25 March 2021
Published: 28 March 2021

1. Introduction

Aerostatic spindles are widely employed in various kinds of precision equipment for their distinct advantages of ultra-high precision, negligible friction, and wide rotating speed range [1]. However, the drawbacks of low capacity, low stiffness, and poor high-speed stability restrict their possible application significantly. Many factors have a great impact on the overall performances of aerostatic spindles, such as the geometrical parameters of the aerostatic bearing, the supply pressure, the rotating speed, and the restrictor type, etc. Hence, it is difficult to optimize the design parameters of aerostatic spindles comprehensively.

Numerous research studies have been conducted to facilitate the design of aerostatic bearings by investigating the impact of the air film geometrical dimensions on their static performance [2,3]. Li et al. [4] numerically studied the performances of the orifice-type restricted aerostatic thrust bearings with varying geometrical parameters based on computation fluid dynamics, and the effect tendency of geometrical parameters to the bearing performances was acquired. Belforte [5] and Du et al. [6] investigated the stiffness, load-carrying capacity (LCC) of aerostatic thrust bearings and journal bearings with pressure-equalizing groove (PEG), and found that the PEG can enhance the static performance of aerostatic bearing dramatically.

Many studies also have been carried out on the optimal design approach of air bearings [7–9]. Wang et al. [10] proposed an optimal design approach based on a genetic algorithm, and a hybrid selection scheme was adopted when selecting the mating groups, which is more efficient. Federico et al. [11,12] optimized the orifice diameter, orifice position, and orifice number of inherently compensated rectangular aerostatic pad bearings based on a multi-objective genetic algorithms optimization approach. To optimize the design parameters and supply pressure of a rectangular aerostatic thrust bearing, Nikhil et al. [13] proposed a Pareto optimization method to avoid the variation of the optimal results with different initial populations; the trial points are therefore distributed in the design space uniformly.

However, the structural deformation caused by pressurized air film was neglected in the aforementioned researches. Since the performance of the aerostatic bearings is significantly impacted by the air film thickness, this phenomenon may induce great error when calculating the performance of the aerostatic, and the initial design requirements may not be satisfied by the actual performance of the aerostatic bearing. This phenomenon is especially dramatic in the design of aerostatic bearings with high stiffness and high LCC, such as the spindle of fly cutting machine tools, rotary tables, etc., and has drawn increasing attention in recent years [14,15]. Lu et al. [16,17] numerically and experimentally investigated the performance of an air spindle considering the fluid–structure interaction (FSI) effect. They found that the LCC curve declines slightly due to the FSI effect, and the axial stiffness declines with the decrease of the thrust thickness plate. In a study by Gao et al. [18], the performance of an aerostatic spindle was investigated by a two-way FSI simulation model. It showed that the air film thickness changes about 7.65 μm due to the structure deformation and the stiffness of the spindle declines about 34%.

According to the above literature review, it can be seen that the design of aerostatic bearing includes the design of the air bearing parameters and the design of the crucial dimensions of the solid parts. Even though many optimal design approaches of air bearings have been proposed, the FSI effect is seldom considered, and the crucial structural dimensions are usually decided empirically. Therefore, the aerostatic spindle cannot be optimized comprehensively from its LCC, volume flow rate (VFR), stiffness, natural frequency, etc.

To overcome this issue, a two-round optimization design method for the aerostatic spindle with consideration of the FSI effect is proposed in the current study. The design parameters for the aerostatic bearing and the solid parts can be optimized by the proposed approach. As an application case, an aerostatic spindle is optimized. The impact of the crucial parameters of air film and the structural dimensions on the performances of the aerostatic spindle are investigated based on the finite element method (FEM). The performance of the aerostatic spindle is enhanced from the aspects of LCC, VFR, stiffness, natural frequency, etc. Furthermore, experiments were implemented to validate the reliability of the calculation results.

2. A Two-Round Optimization Design Method of Aerostatic Spindles Considering the FSI Effect

According to the existing design philosophy of aerostatic bearings [19–21], the design of air bearing is usually first carried out based on its performance requirements and design restrictions, and then the crucial solid parts are designed empirically based on the dimensions of air film. The elastic deformation of crucial parts caused by high air pressure and its impact on the overall performance of the aerostatic spindle are ignored. However, when the FSI effect is considered, the design of the aerostatic spindle consists of the design of the geometrical parameters of air bearing and the design of key structural dimensions. Both aspects impact the overall performance of the spindle system significantly. To facilitate the optimal design of aerostatic spindles considering FSI, a two-round optimization design method is proposed in this research, as shown in Figure 1. Its basic idea is to optimize the geometrical parameters of aerostatic bearing and the key structural dimensions of solid parts sequentially so that the static and dynamic performance of the spindle system can be optimized comprehensively.

In the first-round design stage, the static performances of the aerostatic bearing are improved by optimizing its geometrical parameters. The FEM model of aerostatic bearing is built in this stage. Then, the influence of the crucial design parameters (such as the orifice diameter, the air film thickness, and the orifice number) on the static performance of the aerostatic bearing can be acquired. Finally, the optimized design parameters for air bearing design can be acquired by selecting a good combination of LCC, stiffness, and volume flow rate (VFR).

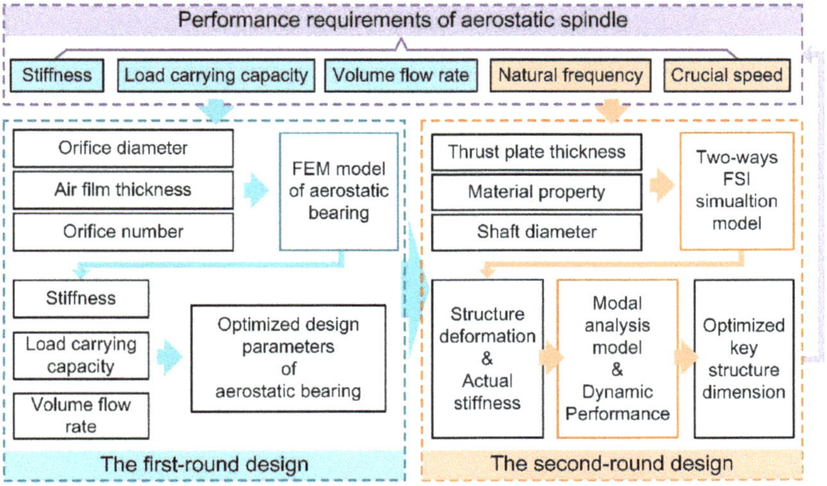

Figure 1. A two-round optimization design method for aerostatic spindles considering the fluid–structure interaction (FSI).

In the second-round design stage, the static and dynamic performances of the aerostatic spindle are further investigated with consideration of the FSI effect. A two-way FSI simulation model is built to calculate the structural deformation of the spindle and its impact on the static performance of the aerostatic bearing. The crucial structural dimensions that greatly influence the structural rigidity and weight of solid components are analyzed. Furthermore, to clarify the effect of structure dimension on the natural frequency of the spindle, modal analysis with varying structure dimensions is carried out, and finally, the optimal structural dimensions are acquired.

In the following sections, the optimal design of an aerostatic spindle is implemented as a case study to elaborate the detailed procedures of the proposed two-round optimization design method.

3. The Configuration of an Aerostatic Spindle

Figure 2a presents the configuration of an aerostatic spindle that is employed in an ultra-precision fly cutting machine tool. Its rotating parts consist of a shaft and two thrust plates. Two air thrust bearings and a radial bearing are adopted to separate the rotating parts from the shaft sleeve and resist the external load. In ultra-precision fly cutting machining, the axial direction of the spindle is the error-sensitive direction. Therefore, the axial vibration and tilt vibration of the rotating parts has a significant impact on the machined surface quality [22]. It indicates that enhancing its axial stiffness and angular stiffness is beneficial to reduce its vibration amplitude and the machined surface quality.

Figure 2b demonstrates the geometrical parameters of the aerostatic thrust bearing. The restrictor of aerostatic thrust bearings is a pocketed orifice-type restrictor. The inner diameter of the thrust bearing is D_1, and the outside diameter of the thrust bearing is D_3. There are n orifices distributes equally along the circumferential direction of the thrust bearing, and the diameter is D_2. d and h are the diameter of the orifice restrictor and the thickness of air film, respectively. d_1 and h_1 are the diameter and the depth of the recess, respectively. Table 1 shows the initial values of these parameters.

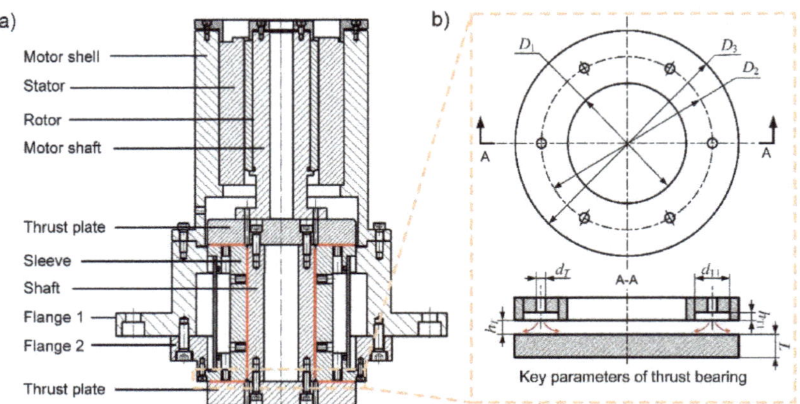

Figure 2. The configuration of an aerostatic spindle and the design parameters of its thrust bearing. (a) The configuration of the aerostatic spindle and (b) the design parameters of the thrust bearing.

Table 1. The design parameters of the aerostatic thrust bearing.

D_1 (mm)	D_2 (mm)	D_3 (mm)	d (mm)	h (μm)	d_1 (mm)	h_1 (mm)	n	L (mm)
60	88.3	130	0.2	16	6	0.05	6	30

4. The First-Round Optimal Design of Aerostatic Thrust Bearings

The static performances of an aerostatic bearing are greatly impacted by the geometrical parameters of the orifice-type restrictor, such as the orifice diameter, the air film thickness, and the orifice number. Therefore, its performances can be improved by optimizing these parameters. In this section, a numerical model of the aerostatic thrust bearing is built first based on the FEM. Additionally, then, the impact of the above three parameters on the static performances of aerostatic thrust bearing is calculated based on the FEM model. Finally, the static performances of the aerostatic thrust bearing are improved by optimizing these three design parameters.

4.1. FEM Modeling of Aerostatic Bearings

The flow status of an orifice restrictor is simplified as the compressible flow through an ideal nozzle, and the orifice is simplified as a point. When the air flows through the orifice, the air pressure drops from P_s to P_d immediately. Considering the difference between the actual mass flow rate of the orifice restrictor and the ideal theoretical mass flow rate, a constant called discharge coefficient $C_d = 0.8$ is introduced to correct the mass flow rate. The mathematical models for calculating the theoretical mass flow rate of the orifice restrictor are Equations (1) and (2) [20].

$$m = C_d A p_s \sqrt{\frac{2\rho_a}{p_a}} \psi_s,\tag{1}$$

$$\psi_s = \begin{cases} \left[\frac{k}{2}\left(\frac{2}{k+1}\right)^{(k+1)/(k-1)}\right]^{1/2}; \left(\frac{p_d}{p_s} \leq \beta_k\right) \\ \left\{\frac{k}{k-1}\left[\left(\frac{p_d}{p_s}\right)^{2/k} - \left(\frac{p_d}{p_s}\right)^{(k+1)/k}\right]\right\}; \left(\frac{p_d}{p_s} > \beta_k\right) \end{cases} \quad \left(\beta_k = (2/(k+1))^{(k+1)/k}\right)\tag{2}$$

where $A = \pi d^2/4$ is the cross-section area of orifice, P_a is the atmosphere pressure, ρ_a is the density of air under standard state, ψ_s is flow rate function, and k is the specific heat ratio of air.

The fluid domain of the aerostatic thrust bearing is described using the governing equation in the cylindrical coordinate system [20] below:

$$\frac{\partial}{\partial \bar{\zeta}}\left(\bar{h}^3 \frac{\partial \overline{p^2}}{\partial \bar{\zeta}}\right) + \frac{\partial}{\partial \theta}\left(\bar{h}^3 \frac{\partial \overline{p^2}}{\partial \theta}\right) + \bar{r}^2 \overline{Q}\delta_i = \Lambda \bar{r}\frac{\partial}{\partial \theta}\left(\bar{h}\bar{p}\right), \tag{3}$$

where $\bar{h}$ is the dimensionless thickness of air film, $\bar{p}$ is the dimensionless pressure, $\bar{\zeta}$ is the dimensionless coordinate in the radial direction, r is the coordinate in the radial direction, and θ is the coordinate in the circumferential direction. $\overline{Q}$ is the mass flow rate factor of the orifice restrictor. δ_i is a constant coefficient with a value of 1 or 0 (in the computational domain where there is an orifice, it equals 1; otherwise, it equals 0.) Λ is the dimensionless bearing number.

$$\bar{h} = \frac{h}{h_m}, \bar{r} = \frac{r}{r_0}, \bar{\zeta} = \ln \bar{r}, \bar{p} = \frac{p}{p_s}, \overline{Q} = \frac{24\eta r_0^2 p_a}{h_m^3 p_s^2 \rho_a}\rho\tilde{v}, \Lambda = \frac{12\eta v_{\theta 2} r_0}{h_m^2 p_s}, \tag{4}$$

where h is the thickness of air film. r_0 is the characteristic length, h_m is the reference air film thickness, p is the gas pressure distribution function, ρ is the air density, η is air dynamic viscosity, $\tilde{v}$ is the velocity of orifice gas flow, and $v_{\theta 2}$ is the velocity components in the circumferential direction. In this paper, the performance of aerostatic thrust bearing is analyzed under motionless conditions, and therefore, $\Lambda = 0$.

The pressure square function is defined as follows:

$$\bar{f} = \overline{p^2}(\bar{\zeta},\theta) = \bar{f}(\bar{\zeta},\theta). \tag{5}$$

Taking the pressure square function $\bar{f}$ as the variable, the functional Φ can be constructed based on Equation (3) as follows:

$$\Phi\left(\bar{f}(\bar{\zeta},\theta)\right) = \int_\Omega \left\{\frac{\bar{h}^3}{2}\left[\left(\frac{\partial \bar{f}(\bar{\zeta},\theta)}{\partial \bar{\zeta}}\right)^2 + \left(\frac{\partial \bar{f}(\bar{\zeta},\theta)}{\partial \theta}\right)^2\right] - \overline{Q}\bar{f}(\bar{\zeta},\theta)\delta_i\right\} d\bar{\zeta}d\theta \tag{6}$$

where Ω denotes the computational domain of the air film. It has been mathematically proved that, under certain boundary conditions, the extremum of Equation (6) can only be acquired by a certain pressure square function $\bar{f}$ that is the solution of Equation (3).

The triangular finite element is employed to solve Equation (6) in this paper. i, j, m are the three nodes of triangular finite elements. The interpolation function of triangular finite element is defined as

$$f = \mathbf{N}^{eT}\mathbf{f}^e \tag{7}$$

$$\mathbf{f}^e = \begin{bmatrix} f_i & f_j & f_m \end{bmatrix}^T \tag{8}$$

$$\mathbf{N}^e = \begin{bmatrix} N_i & N_j & N_m \end{bmatrix}^T, \tag{9}$$

where $\mathbf{f}^e$ represents the pressure square of the nodes. $\mathbf{N}^e$ is the shape function of the triangular element, which can be calculated by Equation (10).

$$\mathbf{N}_i = \frac{1}{2\Delta e}\left(a_i + b_i\theta + c_i\bar{\zeta}\right), (i = i,j,m) \tag{10}$$

$$\begin{cases} a_i = \theta_j\bar{\zeta}_m - \theta_m\theta_j \\ b_i = \bar{\zeta}_j - \bar{\zeta}_m \\ c_i = \theta_m - \theta_j \end{cases}, \tag{11}$$

where Δe is the area of the triangular element. By changing the subscripts of Equation (11) in the order of i, j, m sequentially, and N_j and N_m can be obtained.

By substituting Equation (7) into Equation (6) and applying it to the finite elements of the entire computational domain, Equation (12) can be derived as follows:

$$\Phi(f) = \sum_{e=1}^{m} \frac{1}{2} \int_{\Delta e} \overline{h^3} \left[\left[\frac{\partial}{\partial \overline{x}} \left(\mathbf{N}^{eT} \mathbf{f}^e \right) \right]^2 + \left[\frac{\partial}{\partial \overline{z}} \left(\mathbf{N}^{eT} \mathbf{f}^e \right) \right]^2 \right] d\overline{\zeta} d\theta$$
$$- \sum_{e=1}^{m} \int_{\Delta e} \overline{Q} f d\overline{\zeta} d\theta \delta_i \tag{12}$$

The partial derivative with respect to f should be zero to obtain the solution of Equation (6); thus, we can obtain Equation (13). The solution of Equation (13) is the pressure square function $\overline{f}$, which makes the functional Φ reach the extremum.

$$\frac{\partial \Phi}{\partial f_i} = \sum_{e \in \Delta_i} \left(c_i \mathbf{c}^{eT} + b_i \mathbf{b}^{eT} \right) \mathbf{f}^e \int_{\Delta e} \overline{h^3} d\overline{\zeta} d\theta / (2\Delta e)^2 - k_1 \mu_r \dot{m}_r \delta_i = 0$$
$$(i = 1, 2, \ldots, n) \tag{13}$$

Additionally, Equation (13) can be rewritten in the matrix form as follows:

$$\mathbf{KF} = \mathbf{T}, \tag{14}$$

where $\mathbf{F} = [f_1 \, f_2 \, f_n]^T$ is a column vectors with $n \times 1$ dimensions. $\mathbf{T}$ is composed of three types of elements: 0, $K_1 \dot{m}_r \mu_r$ and $-\left(c_i^{(k)} c_j^{(k)} + b_i^{(k)} b_j^{(k)} \right) \times \int_{\Delta k} \overline{h^3} d\overline{\zeta} d\theta / (2\Delta_k)^2$. $\dot{m}_r$ is the real mass flow rate of the orifice restrictor. $\mathbf{K}$ is a square matrix with $n \times n$ dimensions, and its elements can be acquired by

$$K_{ij} = \sum_{e \in \Delta_i \wedge \Delta_j} \left(c_i c_j + b_i b_j \right) \int \overline{h^3} d\overline{\zeta} d\theta (2\Delta e)^2. \tag{15}$$

Then, the LCC, the stiffness, and the VFR of the aerostatic thrust bearing can be calculated by the equations below. The LCC of the aerostatic thrust bearing is the sum of the integration of the pressure on each element. Then, the stiffness of the aerostatic thrust bearing can be calculated by Equation (17). The VFR can be acquired by Equation (18).

$$W_{LCC} = p_s r_0^2 \sum_{e=1}^{m} \int_{\Delta e} \overline{p} d\overline{\zeta} d\theta \tag{16}$$

$$K_{\text{stiffess}} = \frac{W_{LCC}(h + \Delta h) - W_{LCC}(h)}{\Delta h} \tag{17}$$

$$Q_{\text{VFR}} = m / \rho. \tag{18}$$

Figure 3 presents the FEM model of the aerostatic thrust bearing. Since the thrust air film is symmetric in the circumferential direction, a basic sector is built and the periodic boundary conditions are employed to save calculation time. The computational domain is equally divided into n_θ parts along the circumference direction, while it is divided into n_ζ parts along the radial direction.

4.2. Optimal Design of the Geometrical Parameters of the Restrictor

To optimize the orifice diameter, the air film thickness, and the orifice number of the thrust bearing, their impact on the performance of aerostatic thrust bearing is investigated in this section. Except for these three variables, the values of other geometrical parameters of the air bearing are the same as shown in Table 1.

Figure 4 presents the performance curves of aerostatic thrust bearings with varying orifice diameter d and air film thickness h. It can be seen that with the increase of the air film thickness, the stiffness curve ascents first and then descends, which has a peak. However, the LCC curves decline, and the VFR curve increases monotonically. Moreover, with the increase of the orifice diameter, the LCC and the VFR of the aerostatic thrust bearing increase dramatically. However, for the stiffness curves, the maximum stiffness declines greatly with the increase of orifice diameter. Moreover, the air film thickness, which is corresponding to the peak position of stiffness curves, declines gradually along with the decline of orifice diameter. It indicates that a higher stiffness can be acquired by the combination of smaller orifice diameter and a smaller air film thickness.

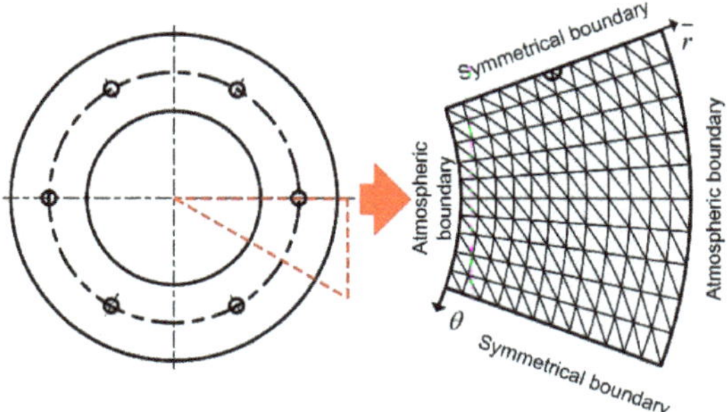

Figure 3. The finite element method (FEM) model of the air thrust bearing.

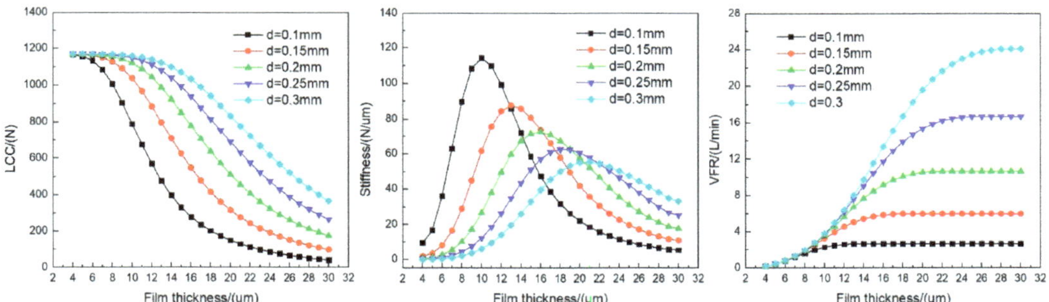

Figure 4. The performance curves of aerostatic thrust bearing with varying d and h.

Figure 5 presents the performance curves of the aerostatic thrust bearing with different orifice numbers n. It shows that the LCC and stiffness of aerostatic thrust bearing can be improved by increasing the orifice number. However, it can also cause a dramatic increase in VFR. In addition, the LCC and stiffness only rise slightly when the orifice number increases to 12, while the VFR increases greatly. This means that the performance of aerostatic thrust bearing cannot be enhanced continuously with increasing orifice number.

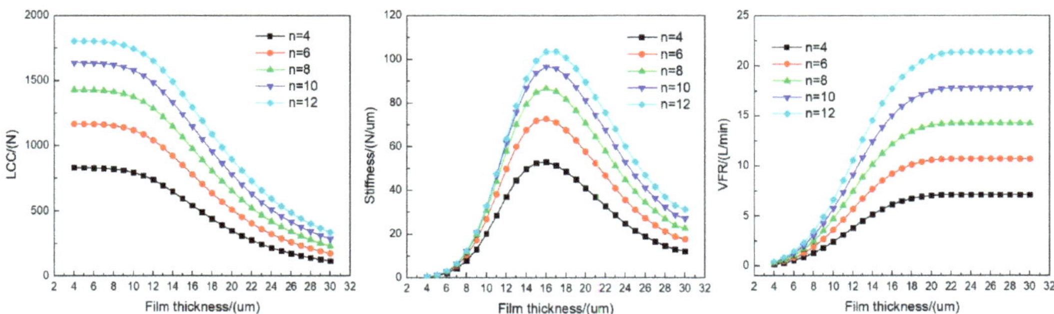

Figure 5. The performance curves of aerostatic thrust bearing with varying orifice number n.

According to the above investigation, the combination of smaller orifice diameter and a smaller air film thickness helps improve the stiffness, and increasing the orifice number can also improve the stiffness. To improve the performance of aerostatic thrust bearing, the orifice diameter of 0.15 mm, the air film thickness of 13 μm, and the orifice number of 8 are selected in this section. Compared with the initial design, the LCC is improved by 28.3%, from 781.0 N to 1001.8 N, the stiffness is improved by 44.2%, from 72.7 N/μm to 104.8 N/μm, and the VFR is reduced by 27.2%, from 9.2 L/min to 6.7 L/min.

5. The Second-Round Optimal Design of Crucial Structural Dimensions

The geometrical parameters of air film determine the design of structural dimensions. After the design parameters of the aerostatic thrust bearing are optimized, the structural dimensions thrust plates are designed empirically based on the dimensions of thrust air film. For example, the outer diameter of air thrust film D_3 should be less than the diameter of thrust plates. However, the thrust plate thickness has a dramatic impact on its bending rigidity and it is generally designed empirically. When the FSI effect is considered, the bending deformation of the thrust plate caused by the air film force could increase the air film thickness and change the stiffness of the aerostatic bearing. Generally, improving the thickness of the thrust plate is beneficial to strengthen its bending rigidity. However, it will also improve the weight of rotating parts, which decreases the natural frequency of the spindle. Therefore, the thrust plate thickness that greatly influences the structural rigidity and weight of rotating parts should be further optimized from the perspective of the natural frequency of the spindle.

5.1. FSI Modeling of Aerostatic Thrust Bearing

To calculate the performance of aerostatic thrust bearing with consideration of elastic deformation of the thrust plate, a two-way FSI modeling method [23] of aerostatic thrust bearing is adopted, as illustrated in Figure 6. It consists of the FEM model of the thrust plate and the FEM model of the aerostatic thrust bearing. The elastic deformation of the thrust plate can be calculated by the FEM model of the thrust bearing. The pressure distribution of the air film can be acquired by the FEM model of the aerostatic thrust film. Since the modeling method of the aerostatic thrust bearing has already been detailed in the previous section, it is not detailed again in this section, and only the modeling method of the thrust plate is explained below.

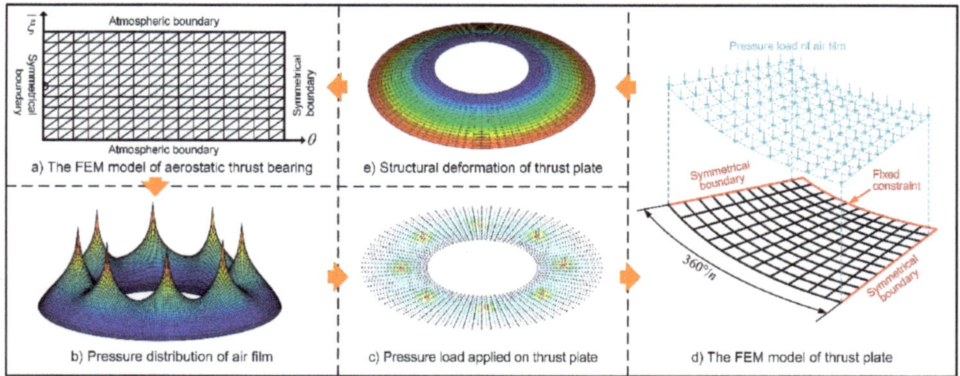

Figure 6. The FSI model of the aerostatic thrust bearing. (**a**) The FEM model of aerostatic thrust bearing; (**b**) the pressure distribution of air film; (**c**) the pressure load applied on thrust plate; (**d**) the FEM model of thrust plate; and (**e**) the structure deformation of the thrust plate.

Considering the structure feature of the thrust plate, a two-dimensional FEM model is built in this section to save calculation time, and the shell element is employed to simulate the bending deformation of the thrust plate. It is usually to simulate the thin-walled to moderately thick-walled structures, which can acquire excellent calculation accuracy and high computational efficiency. As shown in Figure 6d, the inner edge of the thrust plate is constrained by fixed support, and the symmetrical boundary condition is adopted to further save calculation time. The air film pressure is applied on the bearing surface of the thrust plate.

The governing equation for the FEM model of the thrust plate can be written as below.

$$[K]\{u\} = \{F\} \tag{19}$$

where $\{u\}$ and $[K]$ represent the displacement vector and stiffness matrix, respectively. $\{F\}$ donates the pressure load applied on the bearing surface of the thrust plate. According to the Mindin–Reissner shell theory, the stiffness matrix of the shell element can be acquired by Equation (20) [24].

$$\mathbf{K}^e = \frac{h_e^3}{12} \int \int \mathbf{B}_b^T \mathbf{D}_b \mathbf{B}_b dxdy + h_e \int \int \mathbf{B}_s^T \mathbf{D}_s \mathbf{B}_s dxdy \tag{20}$$

where h_e is the thickness of the shell element. $\mathbf{D}$ is the elastic matrix as follows:

$$\mathbf{D}_b = \frac{E}{1-\mu^2} \begin{bmatrix} 1 & \mu & 0 \\ \mu & 1 & 0 \\ 0 & 0 & (1-\mu)/2 \end{bmatrix}, \mathbf{D}_s = \begin{bmatrix} \kappa G & 0 \\ 0 & \kappa G \end{bmatrix} \tag{21}$$

where E denotes the elasticity modulus. μ is the Poisson ratio. κ denotes the shear correction factor. G is the shear modulus. $\mathbf{B}$ can be calculated by Equations (22) and (23).

$$\mathbf{B}_b = \begin{bmatrix} \mathbf{B}_{b1} & \mathbf{B}_{b2} & \cdots & \mathbf{B}_n \end{bmatrix}, \mathbf{B}_s = \begin{bmatrix} \mathbf{B}_{s1} & \mathbf{B}_{s2} & \cdots & \mathbf{B}_{sn} \end{bmatrix} \tag{22}$$

$$\mathbf{B}_{bi} = \begin{bmatrix} 0 & 0 & \frac{\partial N_i}{\partial x} \\ 0 & -\frac{\partial N_i}{\partial y} & 0 \\ 0 & -\frac{\partial N_i}{\partial x} & \frac{\partial N_i}{\partial y} \end{bmatrix}, \mathbf{B}_{si} = \begin{bmatrix} \frac{\partial N_i}{\partial y} & -N_i & 0 \\ \frac{\partial N_i}{\partial x} & 0 & -N_i \end{bmatrix} \tag{23}$$

where N is the shape function of rectangular elements.

$$N_i = (1 + \xi_i\xi)(1 + \eta_i\eta)/4 (i = 1,2,3,4) \tag{24}$$

where ξ_i and η_i are the coordinates of nodes in the local coordinate system of the shell element.

A crucial part of the two-way FSI modeling is to build the bidirectional data exchange interface between the FEM model of the aerostatic thrust bearing and the FEM model of the thrust plate. To realize it, the data exchange procedure is designed and it can be decomposed into the following steps:

(1) Based on the FEM model of the aerostatic thrust bearing, as shown in Figure 6a, conducting a steady-state analysis of air film and the pressure distribution of air film can be acquired, as shown in Figure 6b;

(2) Exporting the air film pressure data to the two-dimensional FEM model of the thrust plate, as shown in Figure 6d, and applying it as pressure load, as shown in Figure 6c;

(3) Conducting steady-state structure analysis based on the two-dimensional FEM model of the thrust plate, in which the deformation of the thrust plate can be acquired, as shown in Figure 6e;

(4) Transferring the deformation data of the thrust plate to the FEM model of the aerostatic thrust bearing and updating the air film thickness of each element;

The bidirectional data exchange can be accomplished by the above procedures. By implementing the above process repeatedly, the static performance of aerostatic thrust bearing can be acquired.

5.2. Optimal Design of the Structural Dimensions of the Aerostatic Spindle

Figure 7 presents the axial stiffness and the first-order natural frequency of the aerostatic spindle with varying thrust plate thickness and air film thickness considering the FSI effect. It can be derived from Figure 7a that with the increase of thrust plate thickness, the maximum stiffness increases sharply in the range of 5–15 mm and then increases slowly in the range of 15–30 mm. This phenomenon can be explained as follows. The structural rigidity of the thrust plate is directly related to its thickness. By increasing its thickness, its structural rigidity will be improved, and the deformation of the thrust plate will be smaller. When the thickness of the thrust plate is larger than 15 mm, the stiffness trends to steady gradually with the increase of the thrust plate. It is because that the structural rigidity of the thrust plate is high enough to resist the air film force. The stiffness is improved slightly by further increasing its thickness. Therefore, the range of 15–30 mm is recommended for the thrust plate thickness for higher stiffness.

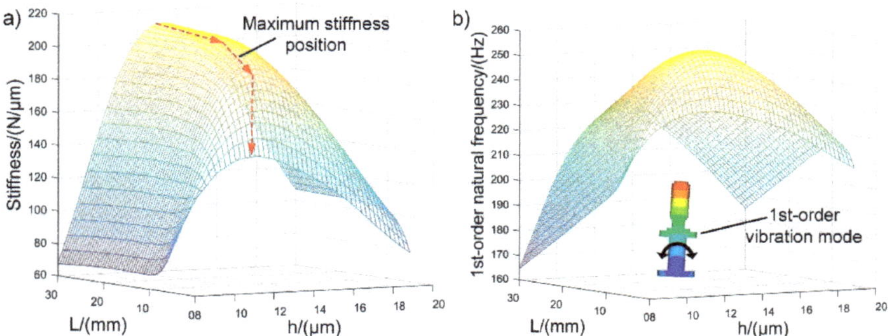

Figure 7. The performance of aerostatic spindle with varying thrust plate thickness *L* and air film thickness *h*. (**a**) The axial stiffness of the spindle with different thrust plate thickness *L* and air film thickness *h* and (**b**) the first-order natural frequency of the aerostatic spindle with varying thrust plate thickness *L* and air film thickness *h*.

However, the mass of the thrust plate is also directly related to its thickness. To increase the natural frequency, the designer generally hopes to reduce the mass of the rotating parts in the design stage. It means that the thickness of the thrust plate cannot be increased blindly. As shown in Figure 7b, with the increase of the thrust plate thickness, the first-order natural frequency first increases sharply and then declines quickly. A maximum region for the first-order natural frequency can be observed around the thrust plate thickness of 10–15 mm. This phenomenon can be explained as follows. According to the calculation result shown in Figure 7a, the stiffness first increases sharply and then trends to steady, while the mass increases linearly with the increase of the thrust plate thickness. Generally, the natural frequency is proportional to the stiffness and is inversely proportional to the mass. In the region where the stiffness increases sharply, the natural frequency increases with the increase of the thrust plate thickness. In the region where the stiffness increases slowly, the natural frequency decreases with the increase of the thrust plate thickness.

Considering the stiffness and natural frequency comprehensively, a thrust plate thickness of 20 mm is selected in this paper. Comparing with the initial design, the first-order natural frequency of the aerostatic spindle increases by 45.3%, from 167.4 Hz to 243.3 Hz.

6. Experimental Validation of the Calculation Result

To validate the above calculation result, the performance of the aerostatic spindle with optimized design parameters is measured in this section. Since the machining error and assembly error is unavoidable, the actual geometry size of the aerostatic bearings may deviate from the design value. The performance of aerostatic bearing is susceptible to orifice diameter and air film thickness. To minimize the errors introduced by the manufacturing and assembly process on the measurement results, the orifice diameter and air film thickness are first measured in this section.

6.1. Measurement of Orifice Diameter

An optical image measuring instrument (Hexagon Optiv Classic 321GL tp) is employed in this research for orifice diameter measuring, as shown in Figure 8a. It has a CCD (Charge Coupled Device) image system to capture the image of the tiny object to be measured, which is capable to measure the contour and dimensions of various complex parts accurately and efficiently. Figure 8b presents a tested image of the orifice. As can be seen from the figure, the round outline of the orifice is clear, and no obvious burr can be observed. These orifices were acquired by micro-drilling machining. The image of the orifice indicates that excellent shape accuracy can be acquired by micro-drilling machining.

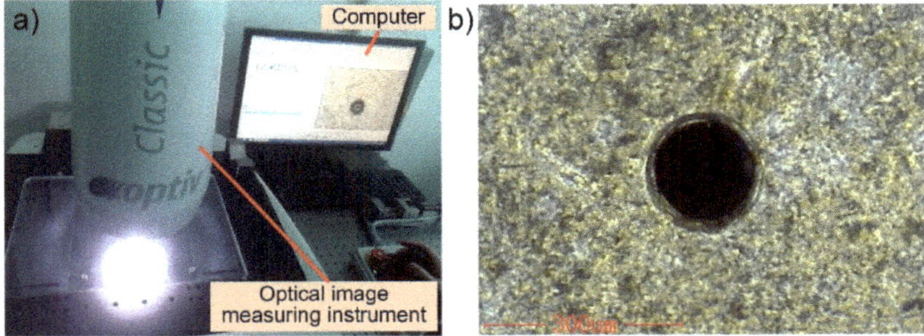

Figure 8. Experimental setup for orifice diameter measuring. (**a**) The Experimental apparatus for orifice diameter test and (**b**) the tested image of an orifice.

Table 2 lists the tested data of eight orifices. It shows that the actual diameter of the orifice diameter has a good consistency. The nominal size of the orifice diameter is 0.15 mm, and all the measured orifice diameter is around 0.15 mm with a dimension deviation less than ±5 μm, which indicates that the orifices have high dimensional accuracy.

Table 2. Experimental result of the orifice diameter.

Orifice Number	1	2	3	4	5	6	7	8
Orifice diameter (mm)	0.148	0.147	0.154	0.148	0.153	0.152	0.151	0.151

6.2. Measurement of Air Film Thickness

Generally, to ensure the static performance of the aerostatic bearing and the motion accuracy of a spindle, a serial of stringent machining accuracy requirements (including flatness, parallelism, and perpendicularity, etc.) are proposed for each part of the spindle system (the shaft, sleeve, and thrust plates). Nevertheless, the real air film thickness may deviate from the design value due to unavoidable machining errors and assembly errors.

Figure 9a presents the experimental setup for air film thickness measuring. By measuring the displacement of the thrust plate under the condition of the air supply valve closing and opening, the actual air film thickness can be acquired. A capacitive displacement sensor (Micro-Epsilon capaNCDT6500) with a resolution of 1 nm and a measurement range of 50 μm is employed in this research to record the displacement of the thrust plate. The experimental setup is mounted on a vibration isolation worktable to avoid the impact of external vibration.

Figure 9b shows the displacement curve of the thrust plate before and after the opening of the air supply valve. It can be seen that the thrust plate moves upward due to the support of pressurized air film. The above measuring process is implemented at eight points. These points are equally spaced along the circumferential direction of the thrust plate, and the tested data is detailed in Table 3. It can be inferred that the measured average air film thickness is about 12.84 µm, which is slightly less than the design value of 13 µm. It may result from the machining error of the shaft, sleeve, and thrust plate.

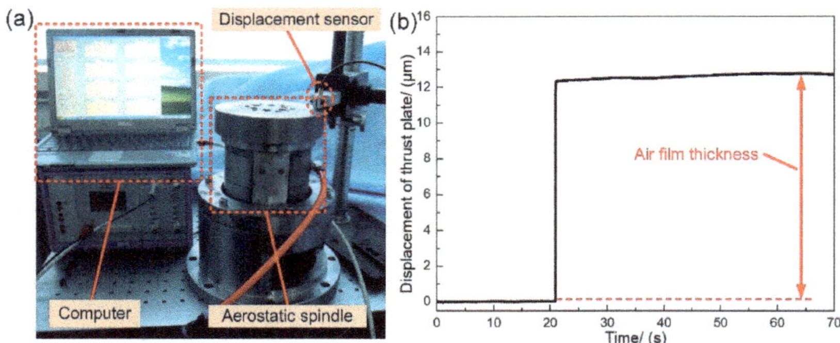

Figure 9. Experimental setup for air film thickness measuring and test result. (**a**) The Experimental setup for air film thickness measuring and (**b**) a test result of the air film thickness test.

Table 3. Measuring result of the air film thickness.

Test Point	1	2	3	4	5	6	7	8
Air film thickness (µm)	12.86	12.72	12.89	12.78	12.93	12.84	12.86	12.84

6.3. The Axial Stiffness Test of the Aerostatic Spindle

To verify the calculation result, the axial stiffness of the spindle with different thrust plate thicknesses is tested in this section. The stiffness test is conducted by applying different loads along the axial direction of the spindle and measuring the axial displacement of the spindle by a displacement sensor. The experimental apparatus mainly consists of a computer, a displacement sensor, and an aerostatic spindle, which is shown in Figure 10. The displacement of the thrust plate can be acquired by the displacement sensor with varying axial loading.

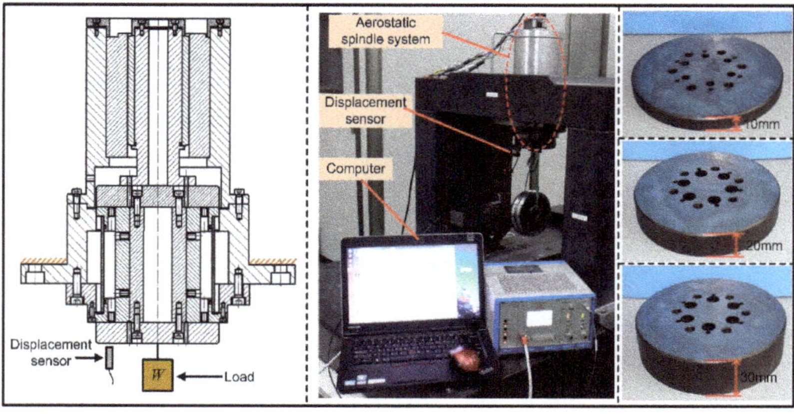

Figure 10. Experimental apparatus for the stiffness test.

The experimental result is compared with the simulation result, as shown in Figure 11. It can be seen that the experimental curve and the simulation curve show the same trend. According to the test date, the axial stiffness of the spindle with the thrust plate thickness of 10 mm is 176.2 N/μm, which is much lower than the ideal stiffness of 209.6 N/μm. It proves that the stiffness will be declined due to structural deformation. By increasing the thrust plate thickness to 20 mm, the stiffness increases to 201.8 N/μm. It indicates that the structural rigidity of the thrust plate can be improved by increasing its thickness. With further increase of the thrust plate thickness to 30 mm, the axial stiffness of the spindle increases slightly to 204.2 N/μm. It means that the stiffness is close to the ideal stiffness with the further increase of the thrust plate thickness. The experimental result validates the accuracy of the FSI model.

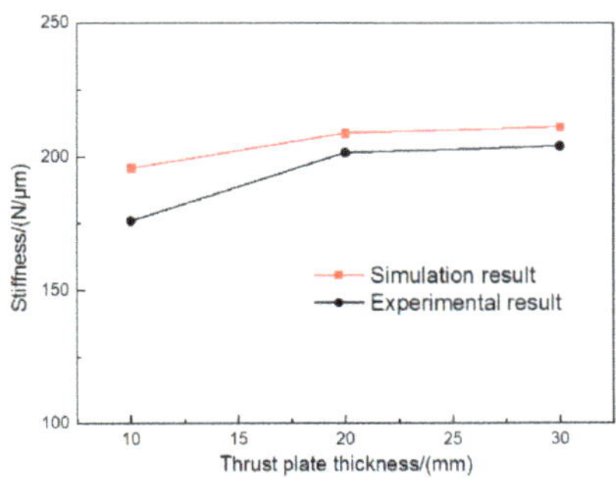

Figure 11. Comparison between the simulation result and experimental result.

7. Conclusions

This paper proposes a two-round optimization design method for aerostatic spindles considering the FSI effect. The optimal design of an aerostatic spindle is implemented as a case study to detail the design process of the proposed approach. A set of optimal design parameters are acquired, and experiments are implemented to verify the simulation results. The main conclusions are drawn as follows:

1. A two-round optimization design method is proposed to facilitate the design of the aerostatic spindle with consideration of the FSI effect, and an aerostatic spindle is optimized as a case study that validates the effectiveness of the proposed design method;

2. In the first-round design stage of the two-round optimization design method, a set of optimized design parameters of the aerostatic thrust bearing is acquired by analyzing the performance of the aerostatic thrust bearing with varying geometrical parameters of the restrictor. Compared with the initial design, the LCC is improved by 28.3%, the stiffness is improved by 44.2%, and the VFR is reduced by 27.2%;

3. In the second-round design stage of the two-round optimization design method, a recommended range for the design of thrust plate thickness is acquired by calculating the stiffness and natural frequency of the spindle with different thrust plate thicknesses. Compared with the initial design, the first-order natural frequency of the aerostatic spindle increases by 45.3%;

4. Experiments are conducted, and the experimental results agree well with the simulation results, which verifies the accuracy of simulation results.

Author Contributions: Conceptualization, Q.G.; methodology, L.L.; software, Q.G. and S.G.; validation, M.Z. and F.Z.; formal analysis, Q.G. and S.G.; investigation, Q.G. and S.G.; data curation, M.Z.; writing—original draft preparation, Q.G.; writing—review and editing, L.L.; visualization, Q.G.; funding acquisition, Q.G. and S.G. All authors have read and agreed to the published version of the manuscript.

Funding: This research was funded by the Heilongjiang Postdoctoral Foundation, grant number LBH-Z19149, the Leading Wild Goose Program Scientific Research Project of Harbin Institute of Technology, grant number XNAUEA5640202220-07, the National Natural Science Foundation of China, grant number 51705501, and the State Key Laboratory of Mechanical Transmissions in Chongqing University, grant number SKLMT-KFKT-201801.

Institutional Review Board Statement: Not applicable.

Informed Consent Statement: Not applicable.

Data Availability Statement: The data presented in this study are available on reasonable request from the corresponding author.

Conflicts of Interest: The authors declare no conflict of interest.

References

1. Gao, Q.; Chen, W.; Lu, L.; Huo, D.; Cheng, K. Aerostatic bearings design and analysis with the application to precision engineering: State-of-the-art and future perspectives. *Tribol. Int.* **2019**, *135*, 1–17. [CrossRef]
2. Chen, Y.; Chiu, C.; Cheng, Y. Influences of operational conditions and geometric parameters on the stiffness of aerostatic journal bearings. *Precis. Eng.* **2010**, *34*, 722–734. [CrossRef]
3. Chen, X.; He, X. The effect of the recess shape on performance analysis of the gas-lubricated bearing in optical lithography. *Tribol. Int.* **2006**, *39*, 1336–1341. [CrossRef]
4. Li, Y.; Ding, H. Influences of the geometrical parameters of aerostatic thrust bearing with pocketed orifice -type restrictor on its performance. *Tribol. Int.* **2007**, *40*, 1120–1126. [CrossRef]
5. Belforte, G.; Colombo, F.; Raparelli, T.; Trivella, A.; Viktorov, V. Comparison between Grooved and Plane Aerostatic Thrust Bearings: Static Performance. *Meccanica* **2011**, *46*, 547–555. [CrossRef]
6. Du, J.; Zhang, G.; Liu, T.; Suet, T. Improvement on load performance of externally pressurized gas journal bearings by opening pressure-equalizing grooves. *Tribol. Int.* **2014**, *73*, 156–166. [CrossRef]
7. Hirani, H.; Suh, N.P. Journal bearing design using multiobjective genetic algorithm and axiomatic design approaches. *Tribol. Int.* **2005**, *38*, 481–491. [CrossRef]
8. Wang, N.; Tsai, C.M.; Cha, K.C. Optimum design of externally pressurized air bearing using Cluster OpenMP. *Tribol. Int.* **2009**, *42*, 1180–1186. [CrossRef]
9. Wang, N.; Cha, K.C. Multi-objective optimization of air bearings using hypercube-dividing method. *Tribol. Int.* **2010**, *43*, 1631–1638. [CrossRef]
10. Wang, N.; Chang, Y.Z. Application of the Genetic Algorithm to the Multi-Objective Optimization of Air Bearings. *Tribol. Lett.* **2004**, *17*, 119–128. [CrossRef]
11. Federico, C.; Marcello, C. Multi-objective optimization of a rectangular air bearing by means of genetic algorithms. *J. Mech. Eng. Autom.* **2012**, *2*, 355–364.
12. Colombo, F.; Della Santa, F.; Pieraccini, S. Multi-Objective Optimisation of an Aerostatic Pad: Design of Position, Number and Diameter of the Supply Holes. *J. Mech.* **2020**, *36*, 347–360. [CrossRef]
13. Bhat, N.; Barrans, S.M. Design and test of a Pareto optimal flat pad aerostatic bearing. *Tribol. Int.* **2008**, *41*, 181–188. [CrossRef]
14. Luo, X.; Han, B.; Chen, X.; Li, X.; Jiang, W. Multi-physics modeling of tunable aerostatic bearing with air gap shape compensation. *Tribol. Int.* **2021**, *153*, 106587. [CrossRef]
15. Gao, Q.; Lu, L.; Zhang, R.; Song, L.; Huo, D.; Wang, G. Investigation on the thermal behavior of an aerostatic spindle system considering multi-physics coupling effect. *Int. J. Adv. Manuf. Technol.* **2019**, *102*, 3813–3823. [CrossRef]
16. Lu, L.; Chen, W.; Yu, N.; Wang, Z.; Chen, G. Aerostatic thrust bearing performances analysis considering the fluid-structure coupling effect. *Proc. IMechE. Part J J. Eng. Tribol.* **2016**, *230*, 1588–1596. [CrossRef]
17. Lu, L.; Gao, Q.; Chen, W.; Liu, L.; Wang, G. Investigation on the fluid–structure interaction effect of an aerostatic spindle and the influence of structural dimensions on its performance. *Proc. IMechE. Part J J. Eng. Tribol.* **2017**, *231*, 1434–1440. [CrossRef]
18. Gao, Q.; Lu, L.; Chen, W.; Liu, L.; Wang, G. A novel modeling method to investigate the performance of aerostatic spindle considering the fluid-structure interaction. *Tribol. Int.* **2017**, *115*, 461–469. [CrossRef]
19. Powell, J.W. *Design of Aerostatic Bearings*; Machinery Pub. Co: Brighton, UK, 1970; pp. 89–99.
20. Liu, T.; Liu, Y.; Chen, S. *Aerostatic Lubrication*; Harbin Institute of Technology Press: Harbin, China, 1990; pp. 51–98.
21. Wang, Y. *Gas Lubrication Theory and Gas Bearing Design*; China Machine Press: Beijing, China, 1999; pp. 144–204.
22. Gao, Q.; Zhao, H.; Lu, L.; Chen, W.; Zhang, F. Investigation on the formation mechanism and controlling method of machined surface topography of ultra-precision flycutting machining. *Int. J. Adv. Manuf. Technol.* **2020**, *106*, 3311–3320. [CrossRef]
23. Gao, Q.; Qi, L.; Gao, S.; Lu, L.; Song, L.; Zhang, F. A FEM based modeling method for analyzing the static performance of aerostatic thrust bearings considering the fluid-structure interaction. *Tribol. Int.* **2021**, *156*, 106849. [CrossRef]
24. Xu, R. *Finite Element Method in Structural Analyses and MATLAB Programming*; China Communication Press: Beijing, China, 2006; pp. 161–188.

Article

Thermal Characteristics Study of the Bump Foil Thrust Gas Bearing

Xiaomin Liu [1,2], Changlin Li [1,2,*], Jianjun Du [1,2,*] and Guodong Nan [1]

[1] School of Mechanical Engineering and Automation, Harbin Institute of Technology, Shenzhen 518055, China; 19B353007@stu.hit.edu.cn (X.L.); hitngd@gmail.com (G.N.)
[2] Shenzhen Key Laboratory of Flexible Printed Electronics Technology, Shenzhen 518055, China
* Correspondence: HITChina_lcl@outlook.com (C.L.); jjdu@hit.edu.cn (J.D.)

Abstract: In this paper, a thermo-hydrodynamic model of the bump foil thrust gas bearing is developed, which solves the coupled gas film three-dimensional energy equation, non-isothermal Reynolds equation, and the foil deformation equation. The effects of bearing speed, thrust load, and external cooling gas on the bearing temperature field are calculated and analyzed. The test rig of foil thrust gas bearing was built to measure the bearing temperature under different working conditions. Both simulation and experiment results show that there exist temperature gradients on the top foil both in the circumferential and radial directions. The simulation results also shows that the top foil side of the gas film has the highest temperature value in the entire lubrication field, and the position of highest temperature moves radially inward on the thrust plate side as the rotor speed increases. The gas film temperature increases with the increasing rotor speed and bearing static load, and rotor speed has greater effects on the temperature variation. Cooling air flow passing through the bump foil is also considered in the simulations, and the cooling efficiency decreases as the mass of gas flow increases.

Keywords: bump foil gas bearing; thermo-hydrodynamic model; temperature field; thermal characteristics analysis

Citation: Liu, X.; Li, C.; Du, J.; Nan, G. Thermal Characteristics Study of the Bump Foil Thrust Gas Bearing. *Appl. Sci.* **2021**, *11*, 4311. https://doi.org/10.3390/app11094311

Academic Editor: Terenziano Raparelli

Received: 8 April 2021
Accepted: 30 April 2021
Published: 10 May 2021

Publisher's Note: MDPI stays neutral with regard to jurisdictional claims in published maps and institutional affiliations.

1. Introduction

Dynamic bump foil gas bearings have been developed since 1960s, and numerous scholars have conducted the researches of related lubrication theories and experiments [1,2]. Nowadays, requirements of larger power density performance are put forward for the turbomachinery, such as blowers, air compressors, and turbochargers, which indicates that gas dynamic thrust bearings need to work under the conditions of higher speeds, larger axial loads, and harsher environments [3]. Compared with traditional gas bearings, bump foil type gas bearings have many significant advantages, such as long operating life, high reliability, large carrying capacity, high speed and high temperature resistance, and good impact resistance. As the rotational speed gradually increases, the elastic foil structure of the bearing deforms, thereby forming the corresponding air film thickness, which has strong adaptability. Therefore, more and more researchers pay attention to the dynamic bump foil gas bearing, which are widely used in a variety of high-speed rotating machinery [4]. Under operating conditions, the ultra-high speed of the rotating thrust plate drives the gas in the clearance to move fast. Due to the viscous friction and gas compression effect between the gas films, the temperature of the bearing will increase sharply. It is insufficient to maintain an ideal temperature inside the bearing only by relying on the bearing self-cooling effect. With the accumulation of high temperature, the air film gap will decrease obviously due to the effect of thermal expansion of the structure, and the wear-resistant coating of the top foil will be damaged, which will eventually cause disastrous damages to the whole rotor system.

In recent years, many impressed studies on the thermal characteristics of gas bearings have been published. Salehi et al. applied the Couette approximation method to solve the temperature field distribution of journal foil gas bearing by coupling the simplified energy equation and the Reynolds equation. However, the calculation model did not consider the heat transfer between the foil structure and rotor [5]. Lee et al. coupled solved the non-isothermal Reynolds equation, gas film thickness equation that considers foil deformation and thermal expansion, and the gas film energy transfer equation. The simulation results of temperature field distribution under the symmetric arrangement of two thrust bearing are obtained, and the influences of rotor speed, bearing load, and cooling air pressure on the bearing temperature are explored [6]. Dickman conducted the thermal-influenced loading experiments based on three corrugated foil thrust bearings with the same size. It was pointed out that the bearing load capacity increases almost linearly with the increase in rotating speed without thermal failure and decreases with the increasing rotor speed when the thermal runaway occurred [7]. GAD et al. calculated the temperature field of thrust bearing by using Couette approximation method, and pointed out that the gas leakage of bearing clearance mostly occurs in the inclined area and leads to the lower temperature. In addition, the study pointed out that more than 70% of the heat generated by gas film can be removed by controlling the cooling flow rate. Meanwhile, the simulation results were compared with Dickman's experimental results, the bearing loading curves in comparison are more consistent when the rotate speed is lower, and there are certain errors when rotor speed is higher. The author speculates that bearing thermal runaway occurred at higher speeds, resulting in a lower load capacity with speed increased [8]. Andreas Lehn et al. present the formulas for calculating the thermal resistances between the bump foil and top foil, and between the bump foil and bearing housing. The authors also considered the thermal expansion of the thrust disc, and the temperature field distribution was solved by multi-field coupling methods. It was pointed out that the bearing load capacity decreased with the increase in rotating speed when the speed increased to a certain value. This trend was due to the bending deformation of thrust disc caused by the high gas film temperature, which leads to the decrease in gas film thickness and the final thermal runaway [9]. Li Changlin established conduction and convection heat transfer models for the bearing sleeve and rotor of the journal bump foil gas bearing, which coupled the non-isothermal Reynolds equations, gas film energy transfer equations, and foil deformation equations. The simulation results compared the THD model with the isothermal model, analyzed the static characteristics and dynamic characteristics of the gas foil bearing [10]. Luo Yixin applied the commercial software to simulate the temperature field of thrust bearing, and studied the influences of different boundary conditions, rotor speeds, inlet pressure, and temperature on bearing thermal characteristics [11]. Peng and Khonsari simplified the foil structure as a line spring model, considered the forced convection cooling outside the flat foil, calculated the multi-dimensional distribution of gas film temperature, but ignored the heat transfer from temperature to rotor and sleeve [12]. Aksoy's research results show that the maximum temperature of the film increases with the grow of load and rotational speed, in which the influence of the rotational speed is greater. At the same time, it is found that the temperature gradient along the axial direction of the rotor can easily cause structural and thermal instability [13]. Andres and Kim found that the viscosity of the gas film increased with the temperature rises, which enhanced the effect of dynamic pressure. The influence of the temperature of the gas film on the structure stiffness of the foil is small. The closer to the shaft end, the greater the gas film temperature gradient, and the axial temperature gradient grew with the increase in the rotational speed [14]. Talmage and Carpino found that the axial temperature gradient will cause the flat foil warping in the end region of the bearing, resulting in uneven axial distribution of the film gap, which makes the flat foil easy to grind with the rotor [15]. Kim and Ki studied the complete heat transfer in a 50KW turbine. It was found that the dynamic stiffness and damping coefficient of the bearing would decrease due to the decrease in the elastic modulus of the foil structure at high temperature [16]. Keun and Andres et al. heated the rotor without cooling gas flow, and the bearing failure occurred when the rotor temperature reached 251 °C [17]. Therefore, the heat control and temperature control strategies, such as injecting cooling gas flow in high temperature environment are very necessary. In this study, a detailed 3D thermal model of bump foil thrust gas bearing is considered. The temperature field distribution of the thrust bearing is solved by coupling the three-dimensional energy equation, non-isothermal Reynolds equation and gas film thickness equation. The cooling models of foil structure and thrust plate, as well as the influence of temperature on the thermal expansion of bearing structure are considered. The thermal characteristics of the bearing under different working conditions are analyzed. Finally, the test rig of the foil thrust gas bearing is built to measure the bearing temperature distributions of bearing under different working conditions.

2. Numerical Simulation Research

The aerodynamic bump foil thrust bearing mainly consists of three parts: the top foil is used to provide lubricating surface; the bump foil plays the role of elastic support; the bearing housing is applied to fix the top foil and the bump foil. The structure of bump foil thrust gas bearing is shown in Figure 1.

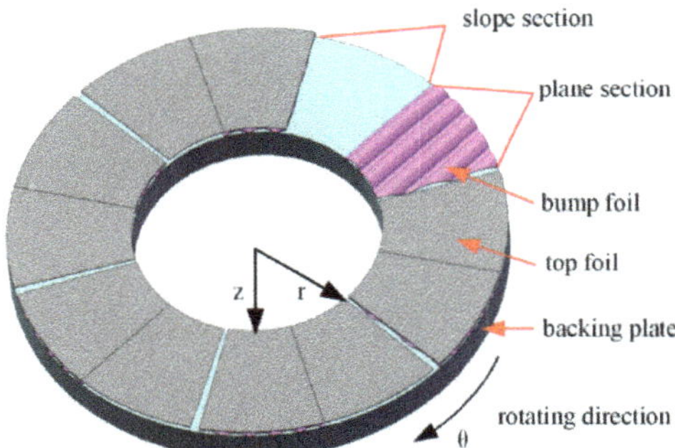

Figure 1. Structure of thrust bump foil gas bearing.

As shown in Figure 2, the entire structural system is divided into six areas along the gas film thickness direction to describe the heat flow. From area A to area G, heat is generated inside the gas film (area C), and one part of the heat is transferred to the ambient or cooling air (area A) through the thrust disc (area B). Another part is transferred to the top foil (area D), and then to the bearing housing (area F) by heat conduction or heat convection (area E) and finally transferred to the open air (area G). When there is cooling air passing through the bump foil channel, only a small part of the heat is transferred to the surrounding air through the bump foil and bearing sleeve.

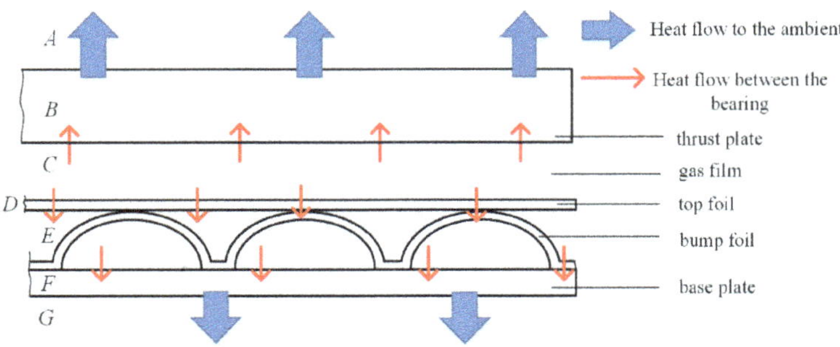

Figure 2. Heat flow transfer model.

2.1. Calculation Process and Parameters

In this paper, the energy equation of the gas film in cylindrical coordinate system is established as follow:

$$\rho(T)c_p(T)\left(v_r\frac{\partial T}{\partial r} + \frac{v_\theta}{r}\frac{\partial T}{\partial \theta} + v_z\frac{\partial T}{\partial z}\right) = k_a(T)\left(\frac{1}{r}\frac{\partial}{\partial r}\left(r\frac{\partial T}{\partial r}\right) + \frac{1}{r^2}\frac{\partial^2 T}{\partial \theta^2} + \frac{\partial^2 T}{\partial z^2}\right) + \left(v_r\frac{\partial p}{\partial r} + \frac{v_\theta}{r}\frac{\partial p}{\partial \theta}\right) + \mu(T)\left(\left(\frac{\partial v_r}{\partial z}\right)^2 + \left(\frac{\partial v_\theta}{\partial z}\right)^2\right) \tag{1}$$

where c_p is the specific heat capacity of gas at constant pressure (J/(kg·K)); k_a is the thermal conductivity of the gas (W/(m·K)); μ is the gas viscosity which changes with temperature, according to the reference [5], the relationship between μ and temperature is expressed as:

$$\mu = a\left(T - T_{ref}\right) \tag{2}$$

where the slope $a = 4 \times 10^{-8}$, when the unit of temperature T is degrees Celsius, $T_{ref} = -458.75$.

The energy equation is nondimensionalized as follow:

$$\overline{\rho}(T)\left(\overline{v}_r\frac{\partial \overline{T}}{\partial \overline{r}} + \frac{\overline{v}_\theta}{\overline{r}}\frac{\partial \overline{T}}{\partial \theta} + \frac{1}{\overline{h}}\overline{v}_z\frac{\partial \overline{T}}{\partial \overline{z}}\right) = k_1\left(\frac{1}{\overline{r}}\frac{\partial \overline{T}}{\partial \overline{r}} + \frac{\partial^2 \overline{T}}{\partial \overline{r}^2} + \frac{1}{\overline{r}^2}\frac{\partial^2 \overline{T}}{\partial \theta^2}\right) + k_2\frac{1}{\overline{h}^2}\frac{\partial^2 \overline{T}}{\partial \overline{z}^2} + k_3\left(\overline{v}_r\frac{\partial \overline{p}}{\partial \overline{r}} + \frac{\overline{v}_\theta}{\overline{r}}\frac{\partial \overline{p}}{\partial \theta}\right) + k_4\frac{1}{\overline{h}^2}\left(\left(\frac{\partial \overline{v}_r}{\partial \overline{z}}\right)^2 + \left(\frac{\partial \overline{v}_\theta}{\partial \overline{z}}\right)^2\right) \tag{3}$$

where k_1, k_2, k_3, k_4 are expressed as follows:

$$k_1 = \frac{k_{air}}{\rho_0 c_p \omega r_2^2} \qquad k_2 = \frac{k_{air}}{\rho_0 c_p \omega \delta_h^2}$$
$$k_3 = \frac{p_a}{\rho_0 c_p\left(T_0 - 273.15 - T_{ref}\right)}$$
$$k_4 = \frac{\mu \omega r_2^2}{\rho_0 c_p\left(T_0 - 273.15 - T_{ref}\right)\delta_h^2} \tag{4}$$

and the dimensionless coefficients in Equation (3) are defined as follows:

$$\overline{\rho} = \frac{\rho}{\rho_0}, \overline{r} = \frac{r}{R_2}, \overline{z} = \frac{z}{h} = \frac{z}{h\delta_h}, \overline{T} = \frac{T - T_0}{T_0 - 273.15 - T_{ref}}$$
$$\overline{p} = \frac{p}{p_a}, \overline{v}_r = \frac{v_r}{\omega r_2}, \overline{v}_\theta = \frac{v_\theta}{\omega r_2}, \overline{v}_z = \frac{v_z}{\delta_h \omega} \tag{5}$$

The velocity components v_r, v_θ, v_z could be solved by the continuity equation of hydrodynamics and the boundary conditions assumptions in Reynold's equation.

2.2. Non-Isothermal Reynolds Equation of Gas Film

When the rotor is running at a higher speed, the heat generated by gas viscous dissipation and compression will increase the gas film temperature. Meanwhile, gas properties such as density and viscosity will change, which can have important impacts on the bearing performance. The non-isothermal gas film Reynolds equation applied in this paper is formulated as:

$$\frac{1}{r}\frac{\partial}{\partial r}\left(rh^3 p\frac{\partial p}{\partial r}\right) + \frac{1}{r^2}\frac{\partial}{\partial \theta}\left(h^3 p\frac{\partial p}{\partial \theta}\right) = 6\mu(T)\omega\frac{\partial(ph)}{\partial \theta} \tag{6}$$

where $\mu(T)$ represents gas viscosity. In the calculation process, the average temperature of the gas film is used to update its viscosity.

2.3. Gas Film Thickness Equation

According to the structure form of the thrust bearing, considering the influence of temperature on the thermal expansion of the bearing structure, the gas film thickness distribution is shown in Figure 3, and its equation is as follows:

$$h = h_2 + g(\theta) + w(r, \theta) - \Delta_{\text{thermal}} \tag{7}$$

$$g(\theta) = \begin{cases} \delta_h[1 - \theta/(b\beta)] & 0 \le \theta < b\beta \\ 0 & b\beta \le \theta \le \beta \end{cases} \tag{8}$$

$$w(r,\theta) = \frac{F(r,\theta)}{K_b \times \Delta l}, \qquad K_b = \frac{Et_b^3}{2(1 - v^2)l^3} \tag{9}$$

where h_2 is the gas film thickness in the plane area (m); $g(\theta)$ is the initial gas film thickness distribution which minuses the value of h_2 (m); β is the angle of top foil (°); b is the pitch ratio; $w(r,\theta)$ is the deformation of top foil under aerodynamic force, and K_b is the stiffness of the bump foil per unit of transverse length (N/m^2) [18]; Δ_{thermal} is the total thermal expansion of all bearing structures (m).

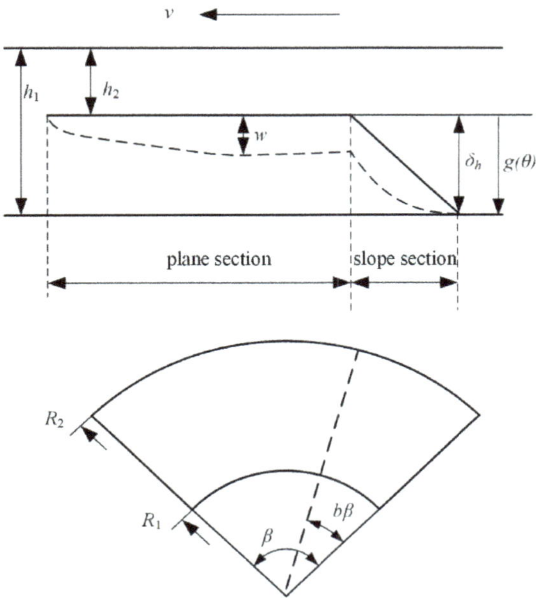

Figure 3. Gas film thickness distribution diagram.

2.4. Bearing Load Capacity and Friction Loss Power

Load capacity ($\overline{F}$) and friction loss power ($\overline{W}$) are the main static characteristics to measure bearing performance, their calculation formulas are as follows:

$$\overline{F} = \frac{F}{p_a r_2^2} = \int_0^\beta \int_{r_1/r_2}^1 (\overline{p} - 1)\overline{r}d\overline{r}d\theta \tag{10}$$

$$\overline{T_r} = \frac{T_r}{p_a h_2 r_2^2} = \int_0^\beta \int_{r_1/r_2}^1 \frac{\overline{r}\overline{h}}{2}\frac{\partial \overline{p}}{\partial \theta} + \frac{\Lambda}{6}\frac{\overline{r}^3}{\overline{h}}d\overline{r}d\theta \tag{11}$$

$$\overline{W} = \frac{2\pi n}{60} \times \overline{T_r} \tag{12}$$

where n is the rotational speed of the thrust plate, $\overline{T_r}$ is friction torque.

2.5. Heat Transfer Model of Bearing Components
2.5.1. Heat Transfer Model of Top Foil Side

In this paper, the equivalent thermal resistance is applied to calculate the overall heat transfer processes when cooling air is introduced or not.

Figure 4 is the heat transfer model on the top foil side. The figure on the left shows that when cold air is not passed into the foil, the heat generated in the gas film passes through the top foil through conduction. Part of the heat is conducted to the base plate through the bump foil, and the other part is conducted to the base plate through the air in the gap between foils. As the base plate is in close contact with the bearing housing, the heat is transferred to the bearing housing through heat conduction, and finally spread out through natural convection. The figure on the right shows that when the cooling air flow is introduced into the foil, the heat generated in the gas film transferred through a different way. The heat passes through the top foil through conduction and part of it is transferred to the bump foil through heat conduction, and another part is taken away by the cold air directly. The heat transferred to the bump foil is conducted to the base plate partly, and the rest of it is taken away by the cold air which passed through the upper and lower surfaces of the bump foil. Then, part of the heat transferred to the base plate is taken away by the cooling air flow on the upper surface of the base plate through convection, the other part is transferred to the bearing housing through heat conduction, and finally spread out through natural convection. Correspondingly, the equivalent thermal resistance models of these two situations are established, respectively.

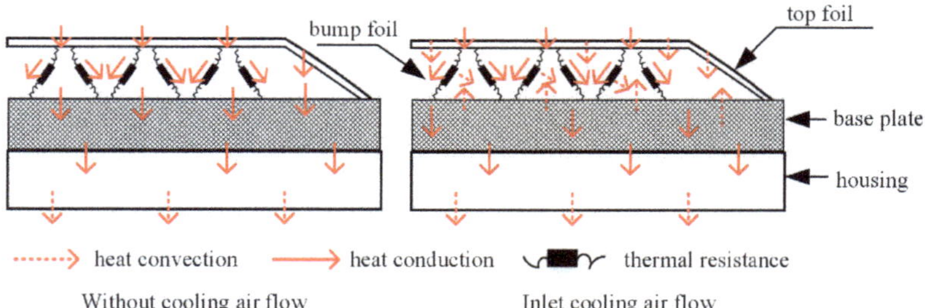

Figure 4. Heat transfer paths near the top foil.

Figure 5 is the schematic diagram of thermal resistance model established based on the ways of heat transfer in the foil structures, where T_1 is the air film temperature on the top foil side; T_0 is the ambient temperature. The remaining parameters are the corresponding equivalent thermal resistance.

When there is no cooling air flow injected, the equivalent total thermal resistance is calculated as follows:

$$R_a = R\_t + \cfrac{1}{\cfrac{1}{R\_bp} + \cfrac{1}{R\_tb}} + R\_b + R\_w + R\_e \tag{13}$$

when the cooling air flow is introduced into the bump foil, the equivalent total thermal resistance is calculated as follows:

$$R_a = R\_t + \cfrac{1}{R\_bp + \cfrac{1}{\cfrac{1}{R\_hbc} + \cfrac{1}{R\_i}} + \cfrac{1}{R\_c}} \tag{14}$$

where R_i is calculated as follows:

$$R\_i = R\_b + \cfrac{1}{\cfrac{1}{R\_hb} + \cfrac{1}{R\_w + R\_e}} \tag{15}$$

the calculation formula for each thermal resistance is as follows:

$$R\_t, R\_bp, R\_tb, R\_b, R\_w = t_i / k_i A_i$$
$$R\_c, R\_hbc, R\_hb, R\_e = 1 / A_j h_j \tag{16}$$

where *i,j* are the serial number of the thermal resistance; *t* is the thickness on heat transmission direction (m); *k* is the thermal conductivity of the material (W/K·m); *h* is the coefficient of heat convection(W/m^2·K); *A* is the heat transmission area (m^2). Under normal circumstances, the parameters *t* and *A* are determined by the geometric features of the components, exceptionally, the A_i of R_bp is contact area between the bump foil and base plate, and the A_j of the R_hbc is approximated obtained by straight inclined segments of the bump foil.

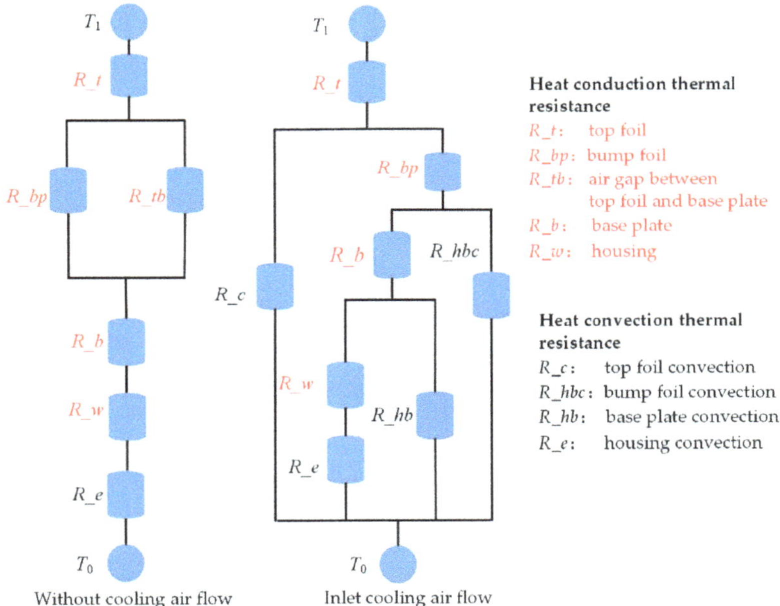

Figure 5. Equivalent thermal resistance model diagram.

When the gas film temperature reaches equilibrium, the heat transferred into the top foil from the gas film will be in dynamic balance with the heat transferred from the top foil side to the outside. Therefore, the heat balance equation between T_1 and T_0 considering the total thermal resistance of the foil structures R_a will form the:

$$-k_a A \frac{\partial T_1}{\partial z} = \frac{T_1 - T_0}{R_a} \tag{17}$$

2.5.2. Heat Transfer Model of the Thrust Plate

Figure 6 is the assembly diagram of a single side thrust bearing. Considering that the thrust plate has high rotating speed, it is assumed that the surface temperature of the thrust plate is consistent along the circumferential direction in the calculation process.

The governing equation of the one-dimensional thrust plate heat conduction model in the radial direction is formulated as:

$$\frac{1}{r} \frac{\partial}{\partial r} \left(k_{disc} r \frac{\partial T_{disc}}{\partial r} \right) + \frac{h_{conv\_disc}}{t_{disc}} (T_{disc} - T_0) + \dot{q}_{film\_disc} + \dot{q}_{disc\_sur} = 0 \tag{18}$$

where k_{disc} is thermal conductivity of thrust plate (W/K·m); T_{disc} is the thrust plate surface temperature (K); h_{conv_disc} is the convection coefficient of air and thrust plate (W/K·m); t_{disc} is the thrust plate thickness (m).

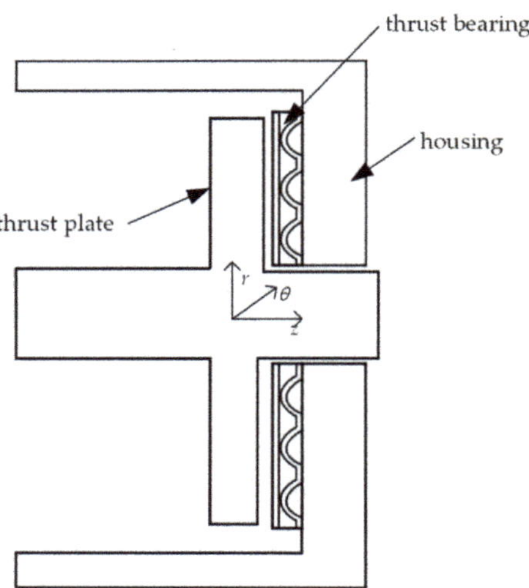

Figure 6. Single side gas thrust bearing configuration.

The last two items represent the heat flux density from the gas film to the thrust plate and from the thrust plate to the ambient environment through the inner and outer radial, respectively. The heat transferred from the air film to the thrust plate is expressed as:

$$\dot{q}_{film\_disc} = \frac{k_{air}}{t_{disc}}\frac{dT}{dz} \tag{19}$$

In this paper, when considering the heat transfer condition on the other side of the thrust plate, the heat dissipation model of the backside of the rotor thrust plate is regarded as the convection heat dissipation of the rotating plate in static air. When calculating the heat dissipation coefficient of convection between the back of the thrust plate and the environment, the calculation formula in reference [19] is used in this article:

$$Nu = 0.015\text{Re}_\omega{}^{0.8} \tag{20}$$

$$h_{conv\_disc} = N_u \times k/L \tag{21}$$

where Nu represents the Nusselt number of the fluid, which is used to describe the convective heat transfer intensity, and Re_ω represents the Reynolds number, and k is the thermal conductivity of thrust plate (W/K·m), and L is the characteristic length (m).

In order the calculate the convective heat dissipation from the thrust plate outer surface to the surrounding air, the convection coefficient on the outer surface of a rotating cylinder is used in this paper [20]:

$$h_{conv\_out} = 0.133\text{Re}_2{}^{2/3}\text{Pr}^{1/3}(k_{air}/D_R) \tag{22}$$

where Re_2 is the Reynolds number with D_R being the characteristic length, and D_R is equal to $2r_2$; r_2 is the outer radius of the thrust bearing (m); Pr is the Prandtl number of air; k_{air} is the thermal conductivity of gas (W/K·m).

$\dot{q}_{\text{disc\_sur}}$ is the heat flux density transferred from the thrust plate to the environment in the radial direction, which is calculated as follows:

$$\dot{q}_{\text{disc\_sur}} = \frac{h_{\text{conv\_out}}}{\Delta r}(T_{\text{disc}} - T_0) \tag{23}$$

2.5.3. Temperature Boundary Condition in the Gas Film Inlet Area

For bump foil thrust gas bearings, the pressure in the gas film area is higher than the ambient atmospheric pressure, thus leading to the leakage of air from the radial boundaries. Meanwhile, the same amount of air Q_{suc} is brought into the gas film at the top foil leading edge, which will mix with recirculation air Q_{rec} to achieve the conservation of the total amount of gas Q_{in}.

The temperature of the inlet mixed air is calculated as follows based on the energy balance condition of the controlled gas volume:

$$T_{in} = \frac{T_{rec}Q_{rec} + T_{suc}Q_{suc}}{Q_{rec} + Q_{suc}} \tag{24}$$

where T_{in} is the mixed gas temperature at the leading edge of the top foil (K); T_{rec} is the circulating air temperature at the rear edge of the top foil (K); T_{suc} is the temperature of the intake cooling air (K).

3. Thermal Analysis

3.1. Calculation Process and Parameters

Figure 7 shows the calculation algorithm of the air-thermoelastic coupling simulation, and Table 1 lists the bearing parameters used in the numerical and experimental studies.

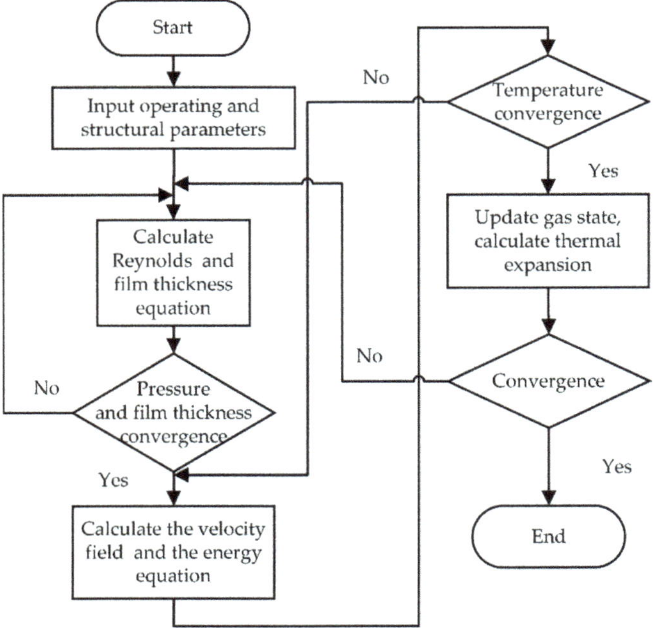

Figure 7. Algorithm of the air-thermoelastic coupling simulation.

Table 1. This is a table. Tables should be placed in the main text near to the first time they are cited.

Parameters	Values
Bearing inner radius r_1 (mm)	16
Bearing outer radius r_2 (mm)	32
Pitch ratio b	0.4
Sector tiles number N	6
Slope height δ_h (µm)	40
Top foil thickness t_d (mm)	0.2
Bump foil thickness t_b, height h_b (mm)	0.15, 0.5
Bump half length l, Bump pitch s (mm)	0.98, 3
Elastic modulus E_b (GPa)	214
Poisson's ratio v_b	0.31
Housing thickness t_k (mm)	16
Foil thermal conductivity k_t (W/K·m)	16.9
Thrust plate thickness t_z (mm)	5

3.2. Analysis of Influence Factors of Bearing Temperature

Figures 8 and 9 present the top foil deformations and the dimensionless gas film pressure distributions when the rotor speed is 50 krpm and the initial minimum gas film thickness is 7 µm. Figure 8a,b show the calculation results of the isothermal model and THD model, respectively. It can be seen that the basic variation trends of the top foil deformations calculated by different models are the same, but the top foil deflection calculated by the THD model is larger. This is because the gas film pressure is larger when the temperature effect is considered. The top foil deformation in the inclined section is larger than that of the plane section due to the absence of the corrugated bump foil underneath. The larger foil stiffness in the plane section helps increase the bearing load capacity.

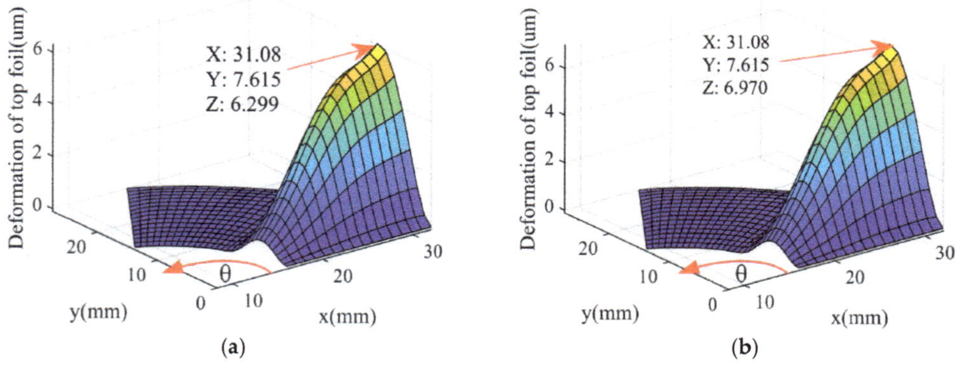

Figure 8. Deformation distribution of top foil under different thermal model. (**a**) Isothermal model; (**b**) THD model.

It can be seen from Figure 9 that the pressure distribution under the two calculation models is basically the same. Along the circumferential direction, the gas film pressure increases sharply from the inlet area, and reaches the maximum value when the gas film transits from the inclined area to the horizontal area. Then it slowly decreases, and finally the air pressure drops to atmospheric pressure in the gas outlet area. In the radial direction, both the inner and outer outlet area are atmospheric pressure, so the pressure presents a trend of increasing first and then decreasing from the inner radius to the outer radius boundaries. There is continuous leakage of gas from the radial boundaries when the bearing is under working condition because of the pressure distribution characteristics. Compared with the isothermal model, the gas film pressure calculated by the THD model is larger than that by the isothermal model. This is due to the fact that the gas viscosity continues to increase as the temperature rises, and the gas film thickness is reduced to a certain extent due to the thermal expansions of the thrust plate and the foil structures.

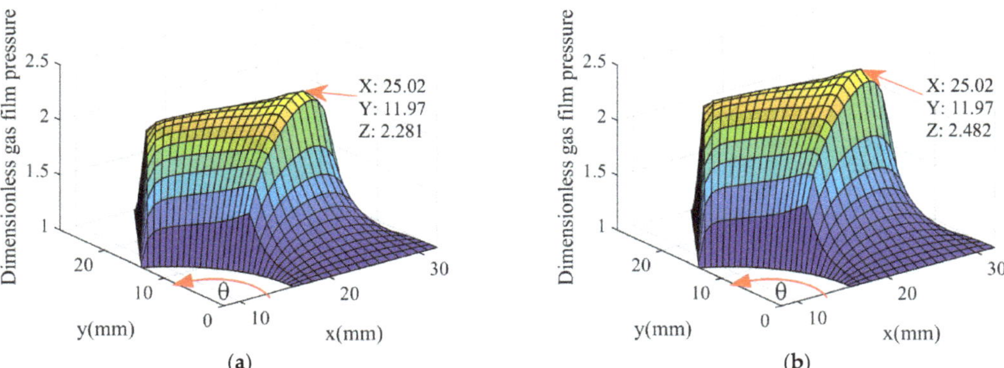

Figure 9. Dimensionless gas film pressure distribution under different thermal model. (**a**) Isothermal model; (**b**) THD model.

Figure 10 shows the change of bearing load capacity and friction loss power with the initial minimum gas film thickness calculated by the isothermal and THD models. It can be seen from the Figure 10a that bearing load capacity decreases with the increase in the initial minimum gas film thickness. When the initial minimum gas film thickness is increased to 30 μm, the gap between the thrust plate and the top foil is too large. Thus, the lubricating gas film is difficult to form, and the bearing almost has no load capacity. It can be seen from Figure 10b that the frictional loss power decreases as the initial minimum gas film thickness increases, which is because the gas film pressure is very small when the initial minimum gas film thickness is large. However, there is no obvious difference between the values of friction loss power calculated by the isothermal model and the THD model. When the initial minimum gas film thickness is the same, the bearing load capacity and friction loss power both increased when the rotor speed goes up, and the difference between the values of the bearing load capacity calculated by the THD model and the isothermal model increased significantly when the speed reach to higher value. This is due to the rapid rise of the gas film temperature caused by the increased rotational speed, and the gas viscosity increases with the temperature rising up.

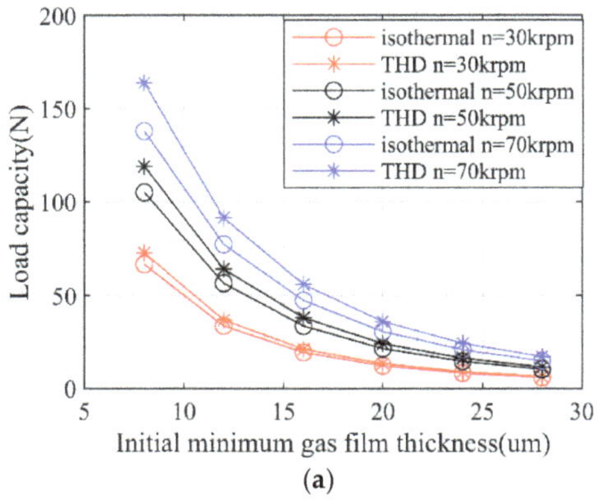
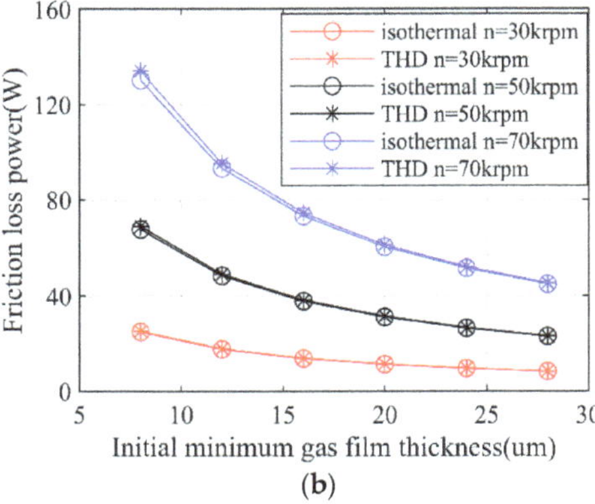

Figure 10. Static characteristics under different thermal model. (**a**) Load capacity; (**b**) Friction loss power.

Figure 11 is the temperature distribution diagram of top foil under the rotor speed of 50 krpm and the static load of 100 N. Similar with the distribution of gas film pressure, the surface temperature of the top foil experiences, firstly, increasing along the rotor rotating direction and then maintaining at a certain value. In the entrance area, the temperature of the ambient gas that is brought into the gas film is relatively low, and it will rise sharply after mixing with the recirculation hot air. The top foil temperature in the plane section is almost the same, which is obviously higher than that in the inclined region. In the radial direction, the temperature increases first and then decreases slightly as the radius increases. This is because the larger radius indicates the larger linear velocity of the gas film and the larger amount of heat generated due to viscous friction. The decreasing trend near the edge can be explained as the heat is dissipated by the forced convection on the thrust plate outer radius surface.

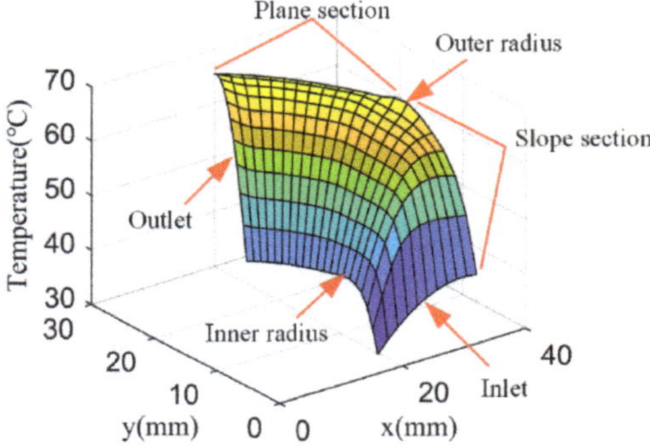

Figure 11. Top foil temperature distribution.

Figure 12 shows the relationship between bearing temperature and the static load at the rotor speed of 50 krpm. The temperatures of top foil and thrust plate are the average values in the simulations. It can be seen from the figure that as the bearing load increases, the gas film temperature increases almost linearly. Additionally, when the bearing load increases, the differences between the maximum air film temperature, the top foil temperature and the thrust plate temperature are almost kept constant. The increase in bearing static load means the gas film is compressed, and the heat generated by the viscous dissipation and the compression in the gas film is larger, resulting in the rise of overall temperature.

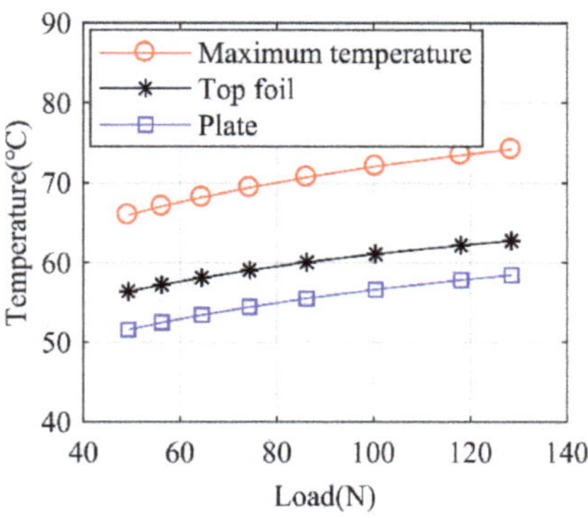

Figure 12. Relationship between bearing temperature and load.

Figure 13 shows the variation of bearing temperature influenced by rotor speed under the bearing static load of 100 N. As the bearing speed increases, the temperatures of gas film, top foil, and thrust plate all increase exponentially. Therefore, rotor speed has a greater influence on the bearing temperature compared with the influence of static load. At low rotor speeds, the difference between the top foil and the thrust plate average temperatures is relatively small, and the maximum temperature of gas film is also relatively small. As the rotational speed increases, the differences between the temperatures of the top foil, thrust plate, and the maximum gas film temperature also increase.

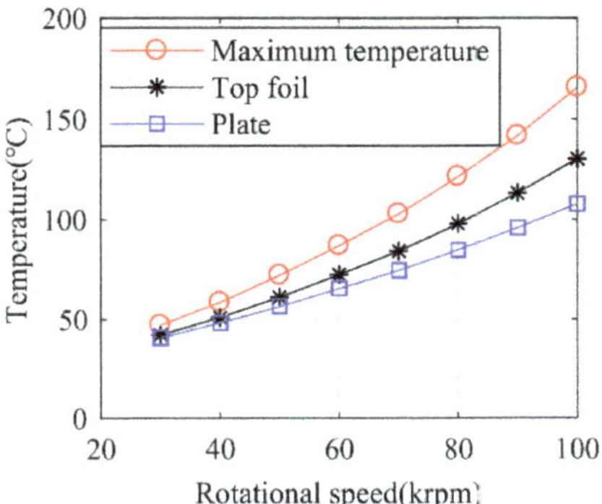

Figure 13. Relationship between bearing temperature and speed.

Figure 14 shows the temperature distributions at various radial positions of the thrust plate at different rotor speeds under the same static load of 100 N. It shows that the temperature of the thrust plate first increases and then decreases in the radial direction, and the highest temperature position is closer to the outer radius boundary. With the increase in rotor speed, the distance from this position to the outer radius boundary tends to become larger.

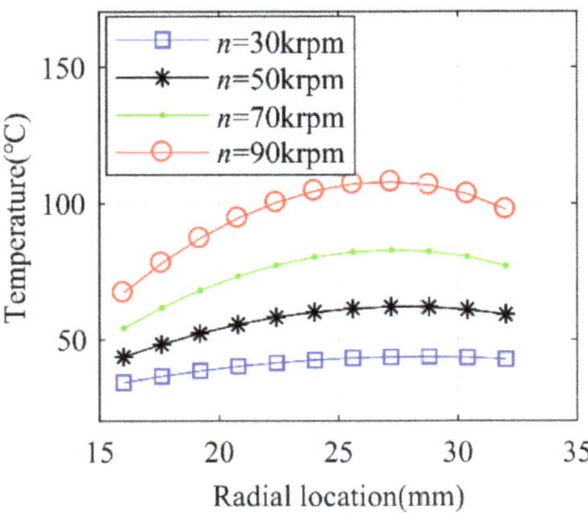

Figure 14. Relationship between thrust plate temperature and speed.

This is because part of the heat is dissipated through forced convection from the cylindrical surface at the thrust plate outer radius, so the gas film temperature will decrease in this area to some extent. As the rotor speed increases, the convective heat exchange effect between the outer edge of the thrust plate and the environment becomes much stronger.

Figure 15 shows the relationship between bearing temperature and the flow rate of cooling gas under the rotor speed of 100 krpm and static load of 100 N. The results show that when the flow rate of cooling gas gradually changes from 0 to 0.2 m^3/min, the bearing temperature drops rapidly. If the gas flow rate is further increased, the cooling effect will be weakened gradually. Therefore, in engineering applications, the appropriate flow of cooling gas passed into the foil structure should be designed to decrease the bearing temperature with the maximum efficiency.

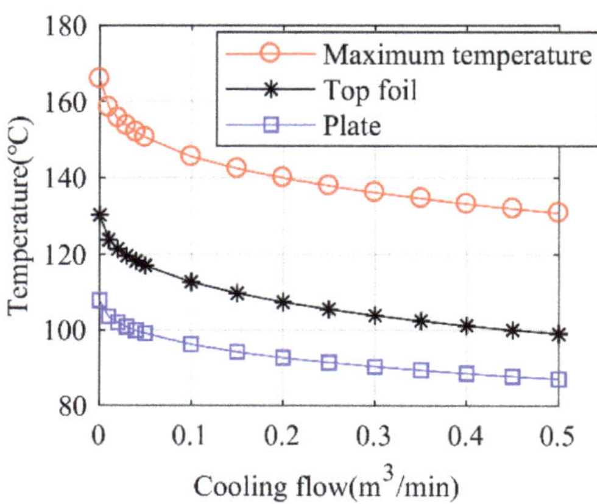

Figure 15. Relationship between bearing temperature and cooling gas flow.

4. Experimental Study

In this paper, the test specimen of the bump foil thrust gas bearing is manufactured and assembled, as shown in Figure 16. The K-type thermocouple is applied to measuring the bearing temperature. A number of thermocouples are stick to the top foil on the back side with copper foil tape. Figure 17 shows the positions of these thermocouples for measuring the local temperature. The thermocouple wire is buried in the bump channel.

The same number and position of thermocouples are pasted under the two centrosymmetric top foils. The temperature measuring points 1–4 are arranged at the radius of 30 mm, and points 5–6 are arranged at the radius of 25 mm. The temperature measuring point 1 is arranged under the slope section, and the remaining temperature measurement points are in the loading region. The assembly of the bearing is completed by a spot-welding machine.

The test bench in this paper is shown in Figure 18. The actual working process of the bearing can be realized by actively controlling the speed of the motorized spindle and by loading the thrustmeter on the other side of the thrust bearing. The displacement between the rotating plate and the bearing is measured by the eddy current displacement sensor.

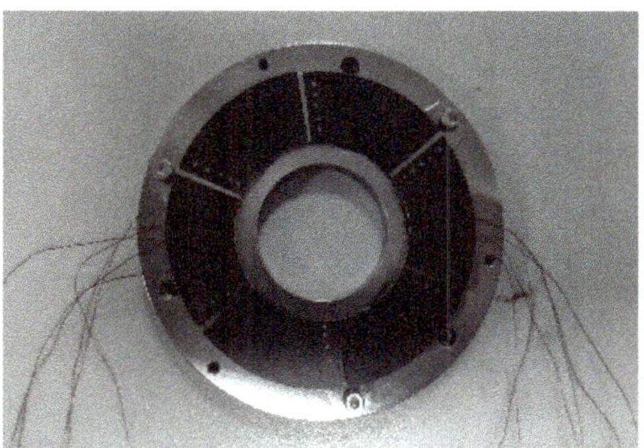

Figure 16. Physical image of experimental bearing.

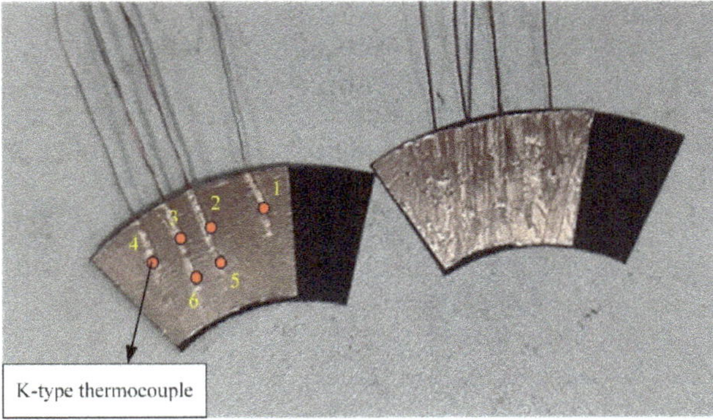

Figure 17. Temperature measuring point bonding.

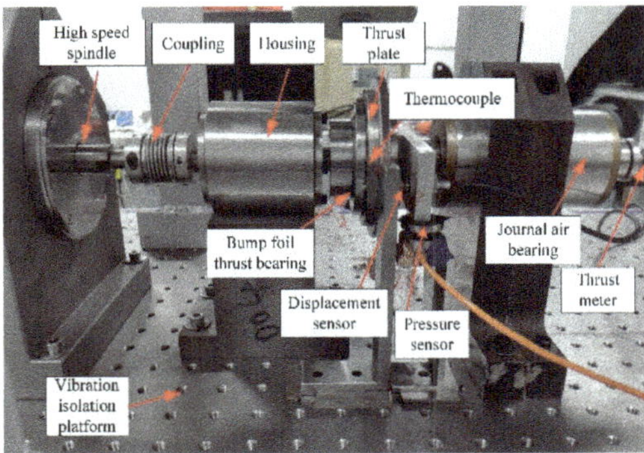

Figure 18. Thrust bearing temperature test bench.

In this experiment, the axial load, which is provided by the SN-500 pointer thruster produced by SUDOO company, is applied on the side of the thrust bearing to simulate the actual axial force of the system. The structure of pointer thruster is shown in Figure 19, where the limit load is 500 N and the resolution is 2 N, the pointer thruster is fixed to the antivibration platform through the supports, and the axial force to the bearing is applied by rotating the button.

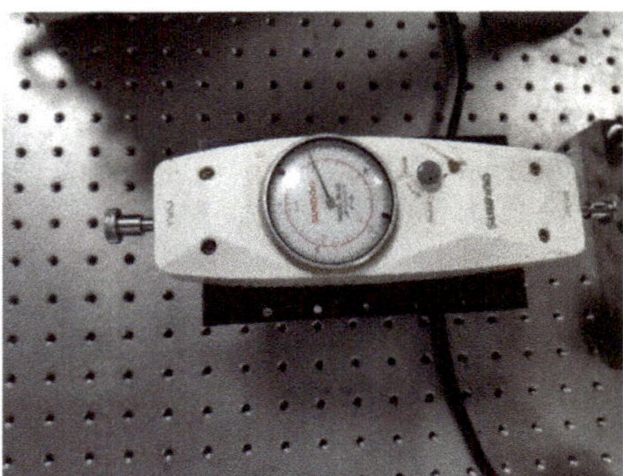

Figure 19. Picture of pointer thruster.

Figure 20 shows the relationship between the tested top foil temperature and the bearing static load at the rotor speed of 15 krpm. During the loading process, the bearing temperature increases continuously. It can be seen from the figure that the measured temperatures on points 2, 3, and 4 are significantly higher than the temperatures on points 5 and 6, and the temperature on point 1 in the slope section is lower compared with the temperatures measured on the points of the same radius. These distribution characteristics of the foil thrust bearing temperature are consistent with the simulation results. Compared with the simulation results, the measured top foil temperatures are relatively lower, which is mainly due to the unclear boundary conditions in the experiments which should be clarified in the simulations.

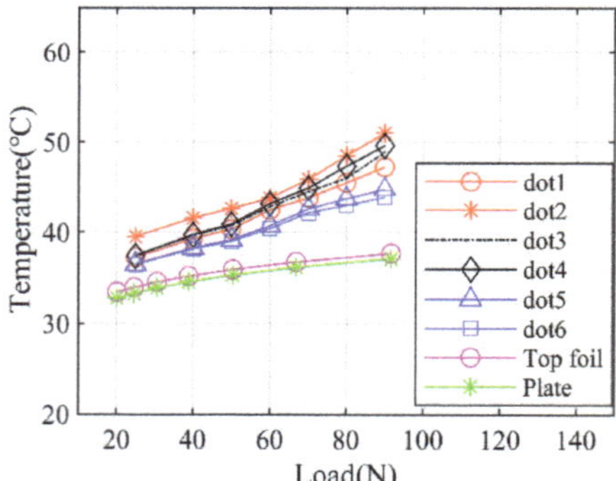

Figure 20. Relationship between bearing top foil temperature and load.

Figure 21 shows the relationship between the measured top foil temperature and rotor speed under the static load of 80 N. It shows that the temperature distribution regularities of the measuring points are similar with the regularities demonstrated in Figure 20, and the overall trend of the measured temperature is the same with the simulation results, which is the temperature will rise up almost linearly with the rotational speed going up. However, the experimental data are relatively higher than the simulation data, which is probably caused by the unclear boundary conditions in the experiments which should be clarified in the simulations or due to the measurement error during the experiment.

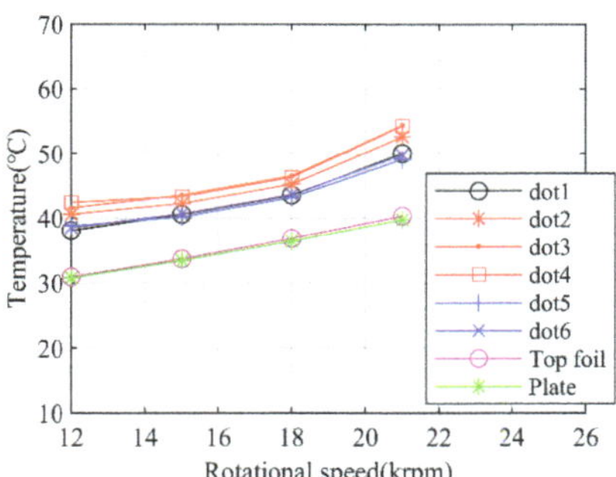

Figure 21. Relationship between bearing top foil temperature and rotational speed.

5. Conclusions

In order to study the thermal characteristics of the aerodynamic bump foil thrust bearing, the air-thermoelastic coupling model is developed in this paper with the heat transfer models of the bearing and thrust plate, and the thermal expansion effect of the bearing structure being considered. The temperature field of the gas thrust bearing is solved through the multi-field coupling algorithm. Finally, the experiments of bearing temperature measurement are carried out under different working conditions. Based on the simulation results, the following conclusions are obtained:

(1) The simulation results show that the surface temperature of the top foil experiences firstly increasing along the rotor rotating direction and then maintaining at a certain value. In the entrance area, the temperature is relatively low, and it will rise sharply after mixing with the recirculation hot air. The top foil temperature in the plane section is almost the same, which is obviously higher than that in the inclined region. In the radial direction, the temperature increases first and then decreases slightly as the radius increases;

(2) The value of bearing load capacity calculated by the THD model is higher than the value calculated by the isothermal model, and the differences of them will become more and more apparent as the rotational speed increased and the initial gas film thickness decreased. Compared to the bearing load capacity, the rotational speed has more effect on the temperature. There is little influence on the friction loss power calculated by different models. Injecting the cooling gas flow under the top foil can reduce the bearing temperature, and its efficiency will decrease with the increase in cooling gas flow. For example, under the situation of and the rotor speed of 100 krpm and static load of 100 N with bearing structure parameters used in this paper, when the gas cooling flow rate of increased up to 0.2 m^3/min, the cooling effect reach to the maximum;

(3) The variation trends of test bearing temperatures with respect to bearing load and rotor speed are similar with those of simulation results, which verified the reasonability of the THD model to some extent. However, the experimental results are relatively higher than the simulation, which is probably caused by the unclear boundary conditions in the experiments which should be clarified in the simulations or due to the measurement error during the experiment.

The current simulation ignored some factors, which makes the simulation results are not exactly the same with the experiment results. The further research will consider more accurate structure model and thermal model, including the surface roughness of the foils and the friction between them, and more complex heat transfer models, etc.

Author Contributions: Conceptualization, J.D. and X.L.; methodology, X.L.; software, X.L.; validation, X.L., C.L., J.D. and G.N.; formal analysis, J.D.; investigation, G.N.; resources, J.D.; data curation, J.D.; writ-ing—original draft preparation, X.L.; writing—review and editing, X.L.; visualization, C.L.;

supervision, C.L.; project administration, J.D.; funding acquisition, C.L. and J.D. All authors have read and agreed to the published version of the manuscript.

Funding: The work described in this paper was supported by National Key R&D Program of China (Grant No. 2018YFB2000100) and the National Natural Science Foundation of China (Grant No. 52005126).

Institutional Review Board Statement: Not applicable.

Informed Consent Statement: Not applicable.

Data Availability Statement: The data presented in this study are available on reasonable request from the corresponding author.

Conflicts of Interest: The authors declare no conflict of interest.

Nomenclature

v_r, v_θ, v_z	Radial, azimuthal, and axial velocity component of the gas flow
ω	Angular velocity of thrust plate
b	Pitch ratio (Ratio of slope section angle to plane section angle)
h	Real gap height in the lubricating gas film
h_2	Prescribed minimal gap height in the lubricating gas film
δ_h	Slope height of the top foil
p	Pressure of the lubricating gas film
p_a	Reference pressure at boundaries
n	The rotational speed of thrust plate
N	Number of pads in the foil thrust bearing
T	Temperature of the lubricating gas film
T_0	Temperature of the enviroment
T_1	Temperature of air film on the top foil side
r_1	Inner radius of the thrust plate
r_2	Outer radius of the thrust plate
td	Thickness of top foil
tb	Thickness of bump foil
tk	Thickness of housing
tz	Thickness of thrust plate
THD	Thermohydrodynamic

References

1. Heshmat, H.; Walowit, J.A.; Pinkus, O. Analysis of Gas-Lubricated Foil Journal Bearings. *J. Lubr. Technol.* **1983**, *105*, 647–655. [CrossRef]
2. Gad, A.M.; Kaneko, S. Performance characteristics of gas-lubricated bump-type foil thrust bearing. *Proc. Inst. Mech. Eng. Part J J. Eng. Tribol.* **2014**, *229*, 746–762. [CrossRef]
3. Dellacorte, C. *Gas Bearing Applications*; Springer: Boston, MA, USA, 2013; pp. 1429–1433.
4. Liu, Z.; Xu, H. Application and Research Progress of Bump-type Foil Aerodynamic Journal Bearings. *Bearings* **2008**, *1*, 39–43.
5. Salehi, M.; Swanson, E.; Heshmat, H. Thermal Features of Compliant Foil Bearings—Theory and Experiments. *J. Tribol.* **2001**, *123*, 566–571. [CrossRef]
6. Lee, D.; Kim, D. Three-Dimensional Thermohydrodynamic Analyses of Rayleigh Step Air Foil Thrust Bearing with Radially Arranged Bump Foils. *Tribol. Trans.* **2011**, *54*, 432–448. [CrossRef]
7. Dickman, J.R. An Investigation of Gas Foil Thrust Bearing Performance and its Influencing Factors. Master's Thesis, Case Western Reserve University, Cleveland, OH, USA, 2010.
8. Gad, A.M.; Kaneko, S. Fluid Flow and Thermal Features of Gas Foil Thrust Bearings at Moderate Operating Temperatures. In *Proceedings of the 9th IFToMM International Conference on Rotor Dynamics*; Springer International Publishing: Cham, Switzerland, 2015.
9. Lehn, A.; Mahner, M.; Schweizer, B. A thermo-elasto-hydrodynamic model for air foil thrust bearings including self-induced convective cooling of the rotor disk and thermal runaway. *Tribol. Int.* **2017**, *119*, 281–298. [CrossRef]
10. Li, C.; Du, J.; Yao, Y. Temperature calculation and static and dynamic characteristics analysis of bump foil gas bearing. *J. Harbin Inst. Technol.* **2017**, *49*, 46–52.
11. Luo, Y.; Zhang, J. Analysis of the aerodynamic and thermal characteristics of the gap of the dynamic pressure gas thrust bearing. *Mach. Build. Autom.* **2019**.
12. Peng, Z.C.; Khonsari, M.M. A Thermohydrodynamic Analysis of Foil Journal Bearings. *J. Tribol.* **2006**, *128*, 534–541. [CrossRef]

13. Aksoy, S.; Aksit, M.F. A fully coupled 3D thermo-elastohydrodynamics model for a bump-type compliant foil journal bearing. *Tribol. Int.* **2015**, *82*, 110–122. [CrossRef]

14. Luis, S.A.; Ho, K.T. Thermohydrodynamic Analysis of Bump Type Gas Foil Bearings: A Model Anchored to Test Data. *J. Eng. Gas Turbines Power* **2010**, *132*, 042504.

15. Talmage, G.; Carpino, M. Thermal Structural Effects in a Gas-Lubricated Foil Journal Bearing. *Tribol. Trans.* **2011**, *54*, 701–713. [CrossRef]

16. Daejong, K.; Jeongpill, K.; Youngcheol, K.; Kookyoung, A. Extended Three-Dimensional Thermo-Hydrodynamic Model of Radial Foil Bearing: Case Studies on Thermal Behaviors and Dynamic Characteristics in Gas Turbine Simulator. *J. Eng. Gas Turbines Power* **2012**, *134*, 052501.

17. Keun, R.; Luis, S.A. On the Failure of a Gas Foil Bearing: High Temperature Operation without Cooling Flow. *J. Eng. Gas Turbines Power* **2013**, *135*, 112506.

18. Heshmat, H.; Walowit, J.A.; Pinkus, O. Analysis of Gas Lubricated Compliant Thrust Bearings. *J. Lubr. Technol.* **1983**, *105*, 638–646. [CrossRef]

19. Trinkl, C.-M.; Bardas, U.; Weyck, A.; aus der Wiesche, S. Experimental study of the convective heat transfer from a rotating disc subjected to forced air streams. *Int. J. Therm. Sci.* **2011**, *50*, 73–80. [CrossRef]

20. Mills, A.F. *Heat Transfer*, 2nd ed.; Prentice Hall: Upper Saddle River, NJ, USA, 1999.

Review

Application of Gas Foil Bearings in China

Yu Hou *, Qi Zhao, Yu Guo, Xionghao Ren, Tianwei Lai and Shuangtao Chen

School of Energy and Power Engineering, Xi'an Jiaotong University, Xi'an 710049, China;
wheredream2012@stu.xjtu.edu.cn (Q.Z.); yu_guo@stu.xjtu.edu.cn (Y.G.); rxh0206@stu.xjtu.edu.cn (X.R.);
laitianwei@mail.xjtu.edu.cn (T.L.); stchen.xjtu@mail.xjtu.edu.cn (S.C.)
* Correspondence: yuhou@mail.xjtu.edu.cn

Abstract: Gas foil bearing has been widely used in high-speed turbo machinery due to its oil-free, wide temperature range, low cost, high adaptability, high stability and environmental friendliness. In this paper, state-of-the-art investigations of gas foil bearings are reviewed, mainly on the development of the high-speed turbo machinery in China. After decades of development, progress has been achieved in the field of gas foil bearing in China. Small-scale applications of gas foil bearing have been realized in a variety of high-speed turbo machinery. The prospects and markets of high-speed turbo machinery are very broad. Various high-speed turbomachines with gas foil bearings have been developed. Due to the different application occasions, higher reliability requirements are imposed on the foil bearing technology. Therefore, its design principle, theory, and manufacturing technology should be adaptive to new application occasions before mass production. Thus, there are still a number of inherent challenges that must be addressed, for example, thermal management, rotor-dynamic stability and wear-resistant coatings.

Keywords: foil bearing; turbo machinery; gas bearing; industrial application

Citation: Hou, Y.; Zhao, Q.; Guo, Y.; Ren, X.; Lai, T.; Chen, S. Application of Gas Foil Bearings in China. *Appl. Sci.* **2021**, *11*, 6210. https://doi.org/10.3390/app11136210

Academic Editors: Terenziano Raparelli, Andrea Trivella, Luigi Lentini and Federico Colombo

Received: 30 May 2021
Accepted: 30 June 2021
Published: 5 July 2021

Publisher's Note: MDPI stays neutral with regard to jurisdictional claims in published maps and institutional affiliations.

1. Introduction

Gas foil bearing has the distinct advantages of cleanliness, low friction, low temperature rise, and cost effectiveness [1]. These advantages make it applicable in most advanced turbo machinery systems, for instance, aerospace, energy and power engineering, etc. [2–4].

Compared with sliding bearing and ball bearing, gas foil bearings have a high running speed and a wider operating range, from tens of thousands rpm (round per minute) to hundreds of thousands rpm. Its power ranges from several kilowatts to more than 100 kilowatts, as shown in Figure 1. It is especially suitable for micro–small high-power density turbo machinery, such as fuel cell air compressors, turbo-expanders, oil-free turbochargers, etc. In comparison, sliding bearings are favored to be used in low-speed applications due to their oil lubricating properties, which are an important support component in large capacity power machinery [5]. As for magnetic bearings, their speed is between that of foil bearings and sliding bearings. Most are used in turbine generator, aeroengine, high-speed CNC (computer numerical control) machine tools, wind power equipment and other cases.

For gas foil bearings, their stability and reliability depend largely on elastic support. Under the elastic top foil, an elastic supporting structure provides additional damping, which conforms to the variable load and speed [6]. Through the damping effect, vibration energy can be absorbed due to Coulomb friction dissipation between the foils and bearing housing. Even if there is impact force or unstable whirl motion, the rotor maintains its stability [7].

In order to improve the bearing performance, different types of gas foil bearings have been proposed. According to the underlying elastic structure, the most-used gas foil bearings are bump foil type, multi-leaf type, protuberant type, and so on [8]. The lubrication mechanisms of a gas foil journal and thrust bearings are basically the same. Herein, a gas foil journal bearing is chosen as an illustration, shown in Figure 2. The basic structure of a bump foil bearing is shown in Figure 2a [9], which consists of a flexible flat foil and an elastic bump foil. The bumps of the supporting foil are evenly distributed along the circumferential direction. The top foil is stacked over corrugated foil, and both foils are fixed at one end and are free at the other end. As for multi-leaf foil bearing, it is composed of a bearing housing and several cantilever foils [10], as shown in Figure 2b. The leading edge of the cantilever foil is fixed in the bearing, while the trailing edge is freely overlapped on the next cantilever foil. The extension direction of the cantilever foil is consistent with the rotation direction of the rotor. A wire can also be used for elastic support. In the wire supporting foil bearing, a certain number of copper wires are fixed on the back of the foil, as shown in Figure 2c [11]. The rigidity of the foil element can be adjusted by changing the number of copper wires and thicknesses of the copper wires. A viscoelastic supporting gas foil bearing is another kind of foil bearing, as shown in Figure 2d. In 1999, the Institute of Refrigeration and Cryogenics of Xi'an Jiaotong University proposed a viscoelastic supporting gas foil bearing [12]. The foil element is a combination of a metal foil and a piece of viscoelastic supporting sheet. The spring supported foil bearing is a new type of foil bearing, as shown in Figure 2e. Hunan University presented a theoretical and experimental study on a novel nested compression spring [13,14]. The stiffness and damping of the support can be adjusted by varying the number and size of springs. In the protuberant foil structure, as shown in Figure 2f, there are multiple layers of protuberant foil between the top foil and the bearing housing [15]. The supporting foil element is composed of a top foil and two protuberant foils. In the assembly process, the bearing performance and stability of the bearing can be improved and the wear on the bearing surface is reduced due to the effective elastic pre-deformation of the protuberant foils [16,17].

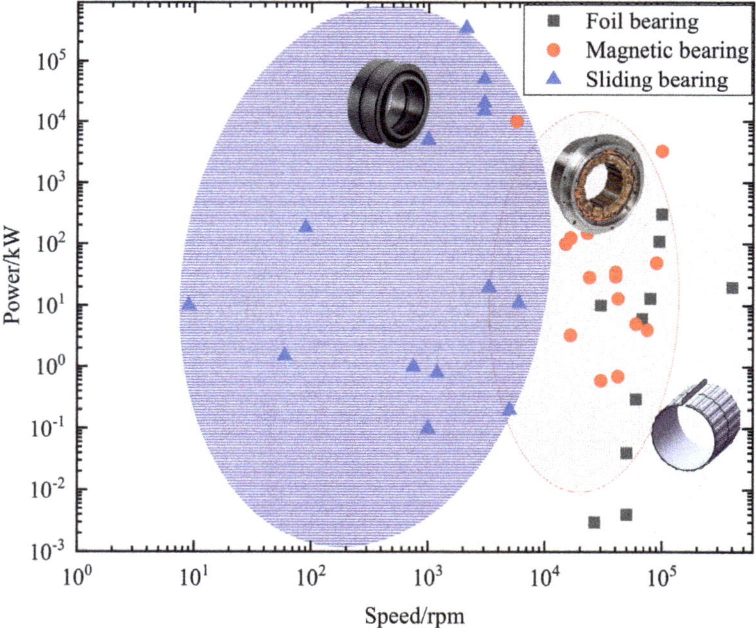

Figure 1. Application range of different kinds bearings.

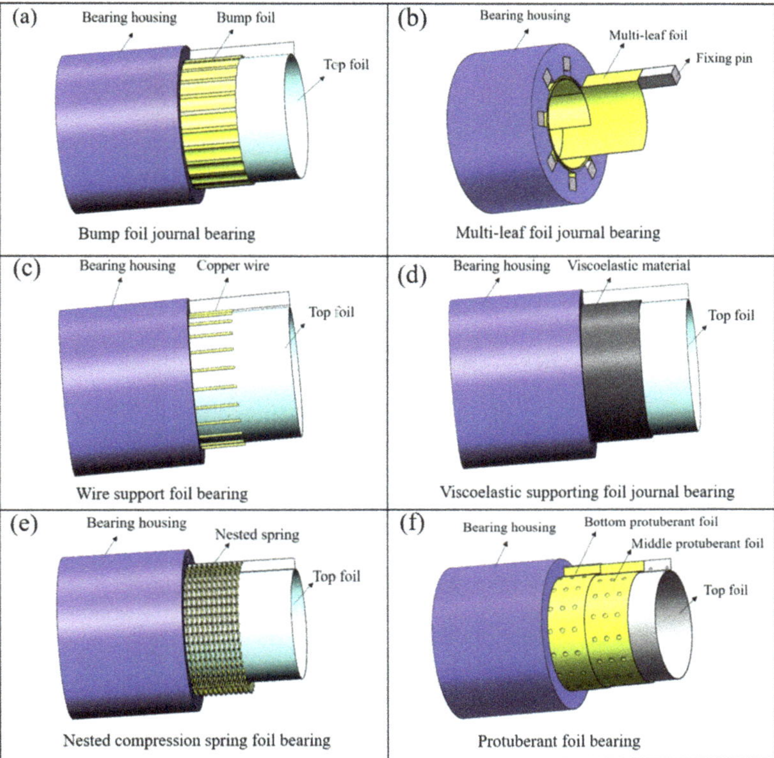

Figure 2. Typical gas foil journal bearings. (**a**) bump foil bearing; (**b**) multi-leaf foil bearing; (**c**) wire supporting foil bearing; (**d**) viscoelastic supporting gas foil bearing; (**e**) spring supported foil bearing; (**f**) protuberant foil bearing.

Before large scale industrial application of the gas foil bearing can be achieved, there are still some practical problems that should be considered. For example, gas foil bearings need to operate stably between takeoff speeds and landing speeds. During starting and stopping processes, the foil will wear poorly in the long run due to the contacting dry friction and the large friction torque. On the other hand, larger power density turbomachinery might lead to a larger axial force, which imposes harsh working requirements on the thrust bearing [18]. In addition, due to the interaction of the foil support structure and the nonlinearity, challenges still exist in accurate theoretical analyses and numerical predictions on foil bearing performance.

The research on gas foil bearings has been carried out relatively early abroad, and mainly used in micro gas turbines [19,20], turbo-expanders [21,22], turbochargers [23–25], fuel cell air compressors machines [26,27], air cycle machines [28], and other options [29]. In recent years, in order to catch up with advanced technology, gas foil bearing technology has been developing rapidly in China. In the 1970s, domestic scholars conducted research on double-row orifice hydrostatic gas journal bearings. Subsequently, a large number of theoretical and experimental studies were carried out in other forms of radial and thrust gas bearings. A series of in-depth studies were conducted in Xi'an Jiaotong University and other institutions, such as, Harbin Institute of Technology, and Beijing Precision Engineering Institute Aircraft industry, etc. [30] For example, a low-temperature turboexpander using rubber-stabilized hydrostatic gas bearings was successfully developed by Xi'an Jiaotong University in 1981. Currently, a series of theoretical and experimental research experiments on gas foil bearings has been conducted at Xi'an Jiaotong University [31],

Hunan University [32,33], the Harbin Institute of Technology [34], Nanjing University of Aeronautics and Astronautics [35–37], the GREE company, and other institutions [38]. Although some progress has been achieved in foil bearings and their applications, a certain number of breakthroughs are still anticipated before large-scale application. In this paper, the progress of domestic gas foil bearings and their applications in high-speed turbo machinery are reviewed, which could provide a reference and guidance for further developments and applications of gas foil bearings.

2. Applications of Gas Foil Bearings

2.1. Application in Low Temperature High-Speed Turbo-Expander

The working temperature of a cryogenic expander is always lower than 120 Kelvin. In order to enlarge the mass flow rate and maintain a relatively high adiabatic efficiency, its rotor needs to run at a high speed under low temperatures. By using the gas foil bearing, the unstable whirl motion of the rotor bearing system can be suppressed and effectively attenuated. Its stability and reliability can be improved notably. In the 1990s, theoretical and experimental research on air foil bearings for turbo-expanders was initiated by the Institute of Refrigeration and Cryogenics of Xi'an Jiaotong University. The rated air volume of the turbo-expander is 600 Nm^3/h@1.15 MPa, and it can work between 0–110 krpm (overspeed to 150 krpm) according to the air supply pressure [39]. The research work on the wire support, viscoelastic support, bump foil support, protuberant support, and multi-leaf foil bearings has been carried out successively, and some progress has been made.

In China, foil bearings were first used in cryogenic turbo-expanders, and then gradually extended to other fields. In 1998, a wire support elastic structure foil bearing was first proposed [40]. In the experiment, a stable speed of 120 krpm or more was reached, and the maximum amplitude was less than 12 μm. The experimental results showed that the turbo-expander ran smoothly at an overspeed of 120% at 118 krpm. In 1999, a turbo-expander supported by viscoelastic foil bearings achieved an excellent performance of 40% over-speeding at 148 krpm [21]. The test used MoS2 powder and two methods of plating Cr and Tic to enhance the hardness of the foil, and over 50 start–stop performance tests were carried out. The test results showed that the surface-treated elastic foil thrust bearing had better start–stop performance and stability. The Institute of Refrigeration and Cryogenics of Xi'an Jiaotong University also conducted a theoretical study on viscoelastic foil bearings [41] and compared the experimental results with wire supported foil bearings. It was found that the stiffness distribution of the viscoelastic bearings was more uniform and had a better stability during ultra-high-speed operations [42]. In 2006, considering the influence of friction, a simplified formula of structural stiffness was obtained, and the structural stiffness of bump foil bearing was analyzed [43]. By comparing two types of foil journal bearings with foil thicknesses of 0.05 mm and 0.07 mm, the vibration characteristics and stability of a turbo-expander were studied. The maximum speed of the turbo-expander using 0.05-mm bump foil bearings was 93,336 rpm. For the turbo-expander using a 0.07-mm bump foil bearing, rotor whirl motion appeared when its maximum speed reached 93,161 rpm. The results showed that when the turbo-expander reached 93 krpm, the maximum amplitude of the rotor was less than 20 μm [44]. In 2012, in order to analyze the feasibility of the application of protuberant foil journal bearings in a turbo-expander, another experimental study was carried out. The results showed that the maximum speed of the turbo-expander reached 99,044 rpm, and the low-frequency whirl motion was small during the entire speed-up and speed-down process [45–47]. In the total protuberant foil bearing test (both journal and thrust bearings are protuberant type), the gas foil bearing ran smoothly and had high repeatability during the operation of the turbo-expander [15]. Subsequently, the multi-leaf foil journal bearing and thrust bearing were simultaneously applied to this high-speed turbo-expander. Transient acceleration and high-speed operation tests of the rotor bearing system were carried out, and the transient characteristics and the stability of the rotor bearing system were analyzed. The experimental results showed that the starting friction torque and running resistance torque of the multi-leaf foil journal bearings were small, and the turbo-expander could reach 93,900 rpm. The rotor bearing system had the advantages of a smaller main frequency amplitude (<6 μm) and a low frequency whirl amplitude (<0.5 μm). The rotor locus was clear and had good repeatability [48,49]. For a cryogenic working fluid turboexpander, as shown in the Figure 3, a helium turbo-expander (500 W) supported by foil journal bearings was designed with a rated speed of 220 krpm. The speed could exceed the design speed by 12% in the test with a normal air temperature. In addition, a series of hydrogen turbo-expanders supported by gas foil journal bearings was also developed, with power ranging from more than 20 to 40 kW. The turbo-expander (39.7 kW) with a rated speed of 74.5 krpm was tested in an air environment. The speed could reach 60 krpm, and the amplitude was less than 0.023 μm. Due to their large thrust force, hydrostatic bearings are currently used as thrust bearings in hydrogen and helium turbo-expanders. Meanwhile, a carbon dioxide turbo expander with a rated speed of 125,000 rpm has been designed, as shown in Figure 4. Its rated speed could reach 80,722 rpm in the experiment. The inlet and outlet pressures of the turbo expander were 10 MPa and 7.3 MPa, respectively.

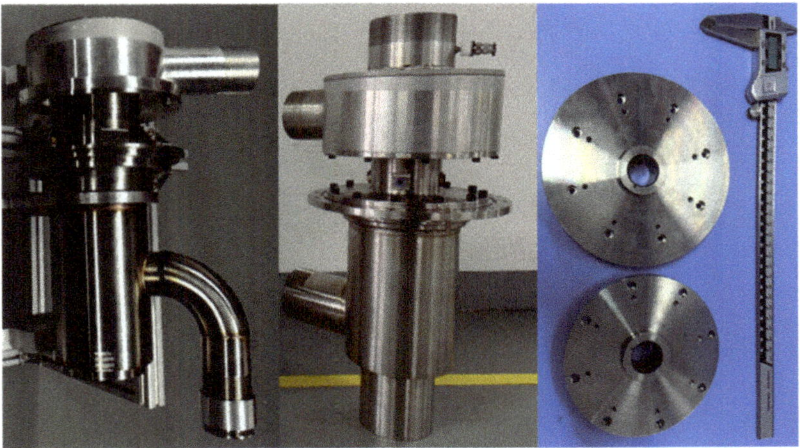

Figure 3. Helium turbo-expander and hydrogen turbo-expander by Xi'an Jiaotong University.

Figure 4. CO_2 turbo-expander by Xi'an Jiaotong University.

2.2. Application in Air Cycle Machine (ACM)

An air cycle machine is always used to adjust the environment in an aircraft cabin, which is the core component of the air conditioning system. It is also widely used in armored vehicles abroad for their high reliability, long service life, and oil-free characteristics. Rotor systems with gas foil bearings have become popular. At present, the main domestic research institutions engaged in ACM research are the Institute of Refrigeration and Cryogenics of Xi'an Jiaotong University, the 16th Research Institute of China Electronics Technology Group Corporation, Hunan University, and other institutions.

Since 2014, domestic research work has been carried out on the ACM system, and some progress has been made. The Institute of Refrigeration and Cryogenics of Xi'an Jiaotong University developed an airborne turbo cooler supported by gas foil bearings [50]. The structure of airborne turbo cooler is shown in Figure 5. The rotor mass is 900 g and the rated speed is 40 krpm. A series of tests have been carried out on its low temperature repeating start–stop and reliability performances. In vibration tests and extreme temperature (−55 °C and 70 °C) start-up tests, its performance was better than expectations. In the tests, the coating did not peel off after a hundred start–stop and over-speed operations. In addition, a 13-kW ACM supported by gas foil bearings was designed and developed by Hunan University, and experimental research was conducted on its rotor bearing system [51]. The critical speed and unbalanced response of the ACM rotor bearing system were analyzed, and the structural parameters of the rotor bearing system were modified according to the analyses. The vibration amplitude of the prototype rotor was finally reduced to about 10 μm. In addition, a performance test on the turbo cooler prototype with foil bearings was conducted by Deng [52]. After 10,000 start–stop life tests and vibration tests, it was found that this machine had a good working performance. So far, the total running hours of this type of turbo cooler exceeded 1000 h.

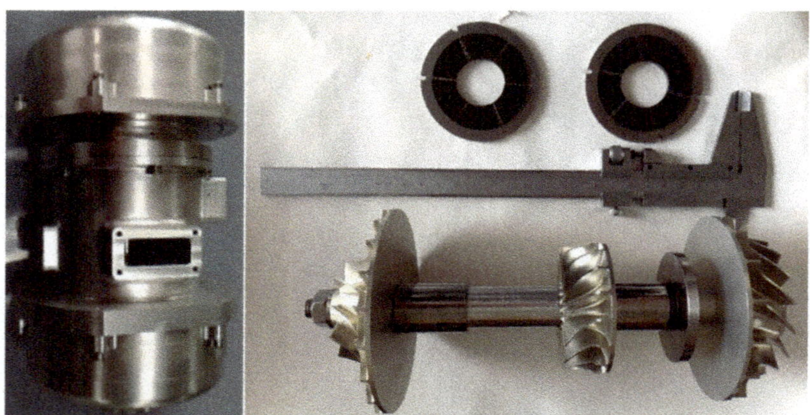

Figure 5. Airborne turbo cooler by Xi'an Jiaotong University [50].

2.3. Application in Centrifugal Blowers

The centrifugal blowers are important in industrial ventilation, which are widely used in many situations, such as medical, chemical, and sewage treatment. Due to the advantages of the high adaptability of the gas foil bearings, the service life and reliability of the blowers are greatly improved under extreme conditions, such as in a dusty environment. On the other hand, blower with gas foil bearings have the merits of low noise, high efficiency, and no pollution. In view of above advantages, centrifugal blowers supported by gas foil bearings have become the first choice for energy-efficient blower. Since the 1970s, foreign blower companies have tried to apply foil bearings in centrifugal blowers. By the end of the last century, highly efficient, energy-saving and highly integrated gas foil bearing blower products have come out on the market. Although domestic theoretical research on gas foil bearings has already been carried out, due to constraints such as materials and process levels, mature products are not yet available on the market. The main domestic research work in this area is the Institute of Machinery Manufacturing Technology, CAEP (China Academy of Engineering Physics) and Shijiazhuang Kingston Bearing Company. Shu X J developed a high-speed air foil bearing with a load capacity of 12.5 kg and tested the bearing performance [53]. By assembling air foil bearings, high-speed motors, and fan impellers, a prototype centrifugal blower was designed and manufactured. The centrifugal blower is shown in Figure 6. The diameter of the rotor is 80 mm and the maximum speed of the original machine is 30 krpm.

Figure 6. Centrifugal blower by the Institute of Machinery Manufacturing Technology, CAEP [53].

2.4. Application in Oil-Free Turbocharger

In the vehicular industry, an engine's output power can be greatly improved by increasing the air–fuel ratio of a turbocharger. Up to now, oil lubricated bearings have mostly been used in automotive turbochargers. Due to the high working temperature, there will be oil quality deterioration and coking, which may lead to catastrophic failure. These problems can be solved effectively by foil bearings. At the same time, the weight, size and complexity of the devices are greatly reduced. In addition, fuel economy and exhaust emissions are greatly improved. At present, domestic studies on turbochargers have also been carried out.

In order to improve turbocharger performance and solve problems, such as oil leakage, Hunan University conducted theoretical and experimental research on oil-free turbochargers supported by air foil bearings, as shown in Figure 7. Subsequent acceleration and deceleration test showed that the rotors exhibited large sub-synchronous vibrations at 20–60 krpm, but the sub synchronous vibration of the rotor was significantly reduced at 68 krpm. The rotor trajectory and FFT analyses showed that the main vibration of the rotor at a steady speed of 68 krpm originated from the synchronous vibration, and the turbocharger could operate stably at 68 krpm [54,55].

Figure 7. Oil-free turbocharger host machine by Hunan University [54].

2.5. Application in High-Speed Gas Compressor

Recently, high-speed compressors supported by gas foil bearings have become an indispensable part of fuel cell systems. Air compressors are mainly used to provide pressurized clean air. The performance of a fuel cell system is directly related to the pressure of the oxygen provided. When the air pressure is increased, the energy density of the fuel cell system can be improved and a higher output power and power performance can be obtained. Up to now, high speed air compressors with gas foil bearings have been put into use in some renewable energy automotive fuel cell systems, for example, Toyota and Honda in Japan, Garrett in the United States, etc.

After nearly ten years of hard work, prototype compressor products can be produced in China. An air compressor with a maximum speed of 120 krpm (10 kW) was developed by the Institute of Refrigeration and Cryogenics of Xi'an Jiaotong University for automotive fuel cell systems. Subsequently, prototype trial production and preliminary tests were completed. The high-speed air compressor is shown in Figure 8. The test results showed that the air compressor operated stably, and the overall technology reached the leading level in China. In addition, the State Key Laboratory for Strength and Vibration of Mechanical Structures (Xi'an Jiaotong University) developed a 11-kW, 70-krpm high-speed experimental air compressor system supported by gas foil bearings. Experiments including loss, rotor transient temperature, and critical speed were carried out and a stable operation of the system under 70 krpm was realized [56,57]. In addition, Hunan University processed and built an oil-free high-speed air compressor experiment test rig [58]. After many test cycles, a diagram of the rotational speed and the temperature characteristics of foil bearings were drawn. Under different experimental conditions, two time-domain waveforms were used to configure the orbit of the rotor system. Meanwhile, the Dalian University of Technology designed a high-speed two-stage air compressor with gas foil bearings [59,60]. Using DyRoBes-Rotor software, the dynamic characteristics of the rotor bearing system were analyzed. The results showed that the rotor-bearing system performances had good repeatability. The maximum bearing force was 60 N, which was much lower than the bearing capacity of 100 N. For refrigerant compressors supported by foil bearings, a centrifugal compressor (630 kW) was produced by the GREE company with a rotor amplitude of less than 12 μm. The unit energy efficiency coefficient of performance (COP) reached 6.35, and the integrated part load value (IPLV) reached 10.15. Meanwhile, the R134a high-speed refrigerant compressor with a rated speed of 80,000 rpm was developed by Xi'an Jiaotong University, and is shown in Figure 8. Its pressure ratio is 2×2, but its size is only one third that of conventional compressors.

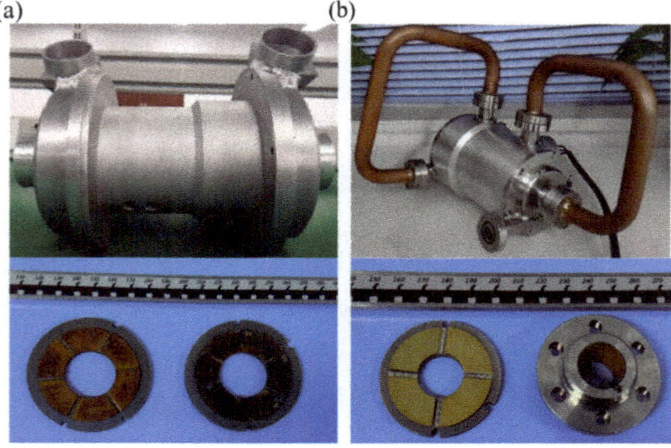

Figure 8. High-speed compressors and gas bearings by Xi'an Jiaotong University (**a**) air compressor; (**b**) refrigerant compressor.

3. Challenges and Prospects

3.1. Load Capacity of Gas Foil Bearing

At present, domestic research on gas foil bearings is mostly focused on gas foil journal bearings, and the research content is mainly on static and dynamic performance, such as bearing capacity and stability. In a rotor bearing system, the foil thrust bearing also plays an equally important role in high-speed turbo machinery. From the perspective of the working principle, there are many similarities between the gas foil journal and trust bearings. Due to the increase in the rotational speed and the power density of turbo machinery, higher requirements are imposed on the bearing capacity and stability of gas foil thrust bearings. Compared with gas foil journal bearings, gas foil thrust bearings require a greater carrying capacity. The bearing capacity and reliability of gas foil thrust bearings have become a bottleneck for turbo machinery. Due to insufficient bearing capacity of the foil thrust bearings, turbo machinery is prone to instability and failure. Research on thrust bearings mainly focuses on experimental research. It is essential to research and develop new high-precision manufacturing and high-performance gas foil thrust bearings. In the development of foil bearing, the operation reliability and the practicability of the manufacturing process also need to be considered. At present, efforts are being made to improve the bearing capacity of journal and thrust foil bearings to 0.6 MPa and 0.4 Mpa, respectively, in China.

3.2. Design Criteria of Gas Foil Bearing

Due to the diversity of industrial applications, higher requirements are put forward for the variety and performance of high-speed turbo machinery. Therefore, design criteria for gas foil bearings and appropriate foil structure parameters are crucial for their actual application; for example, the trade-off relationship between foil stiffness and Coulomb damping. In some cases, a large Coulomb damping is formed due to the high foil flexibility, which is beneficial for improving the stability of the rotor bearing system and lessening the whirl vibration amplitude. However, the large flexibility will cause a large deformation of the foil and rupture the gas film. In addition, the dynamic analysis method of the foil rotor bearing system also needs to be improved. Efforts are being made to increase the operating DN value of the foil bearing to 4.5×10^{6} mm $\cdot$ r/min, in China.

3.3. The Stability of Gas Foil Bearing

The supporting force provided by the gas foil bearings is nonlinear due to the influence of geometric and structural parameters. At the same time, the high-speed rotor system also has relatively complicated dynamic characteristics, which makes stability analysis of the rotor bearing system very complicated. Larger vibration amplitudes are prone to appear in the rotor system. Subsynchronous vibration may cause serious accidents with rotating machinery. At present, a great deal of research has been carried out on subsynchronous vibration of foil bearings. However, the in-depth mechanism of the emergence of subsynchronous vibrations has not yet been found. In order to reveal the influencing mechanism of foil bearing stability, some researches could be carried out from various points of view, such as establishing a complete dynamic solution model considering gas film, foil thickness, and rotor structure, and improving experimental conditions. It is significant to analyze the formation mechanism of the non-linear bearing capacity of the gas foil bearing through experiments and theoretical calculations.

3.4. Thermal Management of Gas Foil Bearing

During high speed operation, a large amount of heat is generated in the foil bearing due to viscous dissipation or high-temperature working environments in some special cases, such as with micro gas turbines. The heat needs to be dissipated or discharged in time, otherwise it will cause a gas foil bearing failure. Therefore, various measures have been taken to transfer heat to prevent heat accumulation, such as processing multi-layer gap foils, adding holes, and adding heat dissipation channels on the bearing housing that can be used to guide cooling air. Apart from this, the development of phase coatings and supporting materials with a high heat capacity is also very beneficial for gas foil bearing thermal management. At present, a great deal of research on the thermal management of foil bearings has been carried out in China. Efforts are being made to increase the maximum operating temperature of foil bearings to 650 °C.

3.5. The Failure Protective Measures and Applications in Extreme Condition

During the operation of the high-speed turbo machinery supported by gas foil bearings, high temperature failure caused by the accumulation of heat and the transient load in the complex environment will cause a sudden stuck of the rotor-bearing system. If the scratches are severe, it will be necessary to replace the rotor with a new one, which would cause a certain amount of economic loss and wasted time. Therefore, failure mechanisms, failure protection, and preventive measures for gas foil bearings need to be considered, which are of great significance to a stable operation. During the operation of turbo machinery, more attention should be paid to the foil temperature and rotor vibration. The temperature of a foil surface can be monitored using temperature sensors in the prevention of bearing failure due to excessively high temperatures. In addition, rotor vibration is an important parameter reflecting the operating stability of high-speed turbomachinery. The vibration displacement, velocity, or acceleration signals can be obtained by data acquisition system. When the rotor amplitude is too large, turbo machinery should be slowed down in time to prevent sudden instability and shutdown. It is a trend to equip foil bearings in aeronautical and astronautical devices. The bearing capacity of foil bearings are greatly affected by working environment, for instance in rarefied air and extreme temperature situations. At the same time, the reliability of foil bearings is required to be higher for aeronautical equipment. The challenge is how to improve the reliability, service life, and load capacity of the foil bearing. Compared with ball bearings and rolling bearings, the load capacity of foil bearings is relatively small. Due to the influence of manufacturing materials, coating technology, and turbine mechanical thermal management, the reliability and service life of foil bearings could be greatly affected.

4. Conclusions

The development of applications for gas foil bearings in China was summarized in this paper. Related challenges during the development process were discussed, including analyses of static and dynamic characterizations, thermal characterization, and lubricating coatings. Based on the above review and discussion, the following conclusions could be drawn:

(1) Gas foil bearing technology developed rapidly in China because of application prospects in high-speed turbo machinery. The diversity of gas foil bearing applications also put forward higher requirements on bearing design. Currently, the design criteria of gas foil bearings need to be configured for different application considerations, especially for high-performance gas foil journal and thrust bearings.

(2) Advances in foil coating technology could greatly improve the reliability and service life of turbo machinery. Research on solid lubricating coating of gas foil bearings was closely related to solid lubricating technology and coating technology. Limited by the temperature range of coating materials, friction pair selection and feasibility of manufacturing and processing, further foil coating technology research is thus needed for bearing applications.

(3) One of the major failure modes for gas foil bearing was due to thermal instability. Research is needed to develop suitable predictive tools to address this problem at the design stage. A detailed non-linear thermo-elastohydrodynamic analysis is needed, considering thermal, hydrodynamic, and structural deformation. In the design and application process, different measures should be adopted to reduce heat accumulation and bearing failure during bearing operation.

Funding: This project was funded by the National Key R&D Program of China (2018YFB2000100), the Na-tional Natural Science Foundation of China (51976150) and the Youth Innovation Team of Shaanxi Universities.

Institutional Review Board Statement: This paper does not involve humans or animals.

Informed Consent Statement: This paper does not involve humans or animals.

Data Availability Statement: The study did not report any data.

Conflicts of Interest: The authors declare no conflict of interest.

References

1. Samanta, P.; Murmu, N.C.; Khonsari, M.M. The evolution of foil bearing technology. *Tribol. Int.* **2019**, *135*, 305–323. [CrossRef]
2. San Andrés, L.; Ryu, K.; Kim, T.H. Thermal Management and Rotordynamic Performance of a Hot Rotor-Gas Foil Bearings System—Part I: Measurements. *J. Eng. Gas Turbines Power* **2011**, *133*, 10. [CrossRef]
3. Emerson, T. Application of Foil Air Bearing Turbomachinery in Aircraft Enviormental Contral Systems. *Control Syst.* **1978**, *100*, 111. [CrossRef]
4. Agrawal, G.L. *Foil Gas Bearings for Turbomachinery*; SAE Technical Paper; SAE International: Hamilton, OH, USA, 1990. [CrossRef]
5. Mokhtar, M.O.A. Floating ring journal bearings: Theory, design and optimization. *Tribol. Int.* **1981**, *14*, 113–120. [CrossRef]
6. Le Lez, S.; Arghir, M.; Frene, J. Static and Dynamic Characterization of a Bump-Type Foil Bearing Structure. *J. Tribol.* **2006**, *129*, 75–83. [CrossRef]
7. Lee, Y.B.; Kim, T.H.; Kim, C.H.; Lee, N.S.; Choi, D.H. Dynamic characteristics of a flexible rotor system supported by a viscoelastic foil bearing (VEFB). *Tribol. Int.* **2004**, *37*, 679–687. [CrossRef]
8. Agrawal, G. Foil Air/Gas Bearing Technology—An Overview. *Am. Soc. Mech. Eng.* **1997**. [CrossRef]
9. Le Lez, S.B.; Arghir, M.; Frene, J. A New Bump-Type Foil Bearing Structure Analytical Model. In Proceedings of the ASME Turbo Expo 2007, Power for Land, Sea, and Air, Montreal, QC, Canada, 14–17 May 2007; pp. 747–757.
10. Oh, K.P.; Rohde, S.M.A. Theoretical Investigation of the Multileaf Journal Bearing. *J. Appl. Mech.* **1976**, *43*, 237. [CrossRef]
11. Wang, L.Z.; Hou, Y.; Cui, M.X.; Chen, C.Z. Damping Evaluation of Foil Bearings. *J. Xi'an Jiaotong Univ.* **2006**, *40*, 40–44. (In Chinese)
12. Chen, R.G.; Zhou, Q.; Hou, Y. Studies of self-acting journal foil bearing for miniature high speed turbo-expander. *Lubr. Eng.* **2010**, *10*, 13–17. (In Chinese)
13. Feng, K.; Liu, W.; Yu, R.; Zhang, Z.M. Analysis and experimental study on a novel gas foil bearing with nested compression springs. *Tribol. Int.* **2017**, *107*, 65–76. [CrossRef]
14. Song, J.-H.; Kim, D. Foil Gas Bearing with Compression Springs: Analyses and Experiments. *J. Tribol.* **2007**, *129*, 628–639. [CrossRef]
15. Lai, T.; Guo, Y.; Zhao, Q.; Wang, Y.; Zhang, X.Q.; Hou, Y. Numerical and experimental studies on stability of cryogenic turbo-expander with protuberant foil gas bearings. *Cryogenics* **2018**, *96*, 62–74. [CrossRef]
16. Lai, T.W.; Ma, B.; Zheng, Y.Q.; Hou, Y.; Zheng, Y.Q. Experiment investigation on stability of multi-decked protuberant foil gas journal bearings. *J. Xi'an Jiaotong Univ.* **2014**, *3*, 79–83. (In Chinese)
17. Wang, W.; Li, X.J.; Zeng, Q.; Lai, T.W.; Hou, Y. Stability analysis for fully hydraodynamic gas-lubricated protuberant foil bearings in high speed turbomachinery. *J. Xi'an Jiaotong Univ.* **2017**, *8*, 84–89. (In Chinese)
18. Łagodzinski, J.; Miazga, K.; Filipiak, I. Application of a compliant foil bearing for the thrust force estimation in the single stage radial blower. *Open Eng.* **2015**, *5*, 105–114.
19. Kozanecka, D.; Kozanecki, Z.; Tkacz, E.; Łagodziński, J. Experimental research of oil-free support systems to predict the high-speed rotor bearing dynamics. *Int. J. Dyn. Control* **2015**, *3*, 9–16. [CrossRef]
20. Lubell, D.; Dellacorte, C.; Stanford, M.; ASME. *Test Evolution and Oil-Free Engine Experience of a High Temperature Foil Air Bearing Coating*; American Society of Mechanical Engineers: New York, NY, USA, 2006.
21. Hou, Y.; Xiong, Y.; Chen, C.Z. Experimental study of a new compliant foil air bearing with elastic support. *Tribol. Trans.* **2004**, *47*, 308–311. [CrossRef]
22. Xiong, L.Y.; Wu, G.; Hou, Y.; Liu, L.Q.; Ling, M.F.; Chen, C.Z. Development of aerodynamic foil journal bearings for a high speed cryogenic turboexpander. *Cryogenics* **1997**, *37*, 221–230. [CrossRef]
23. Sim, K.; Lee, Y.-B.; Kim, T.H. Rotordynamic Analysis of an Oil-Free Turbocharger Supported on Lobed Gas Foil Bearings—Predictions versus Test Data. *Tribol. Trans.* **2014**, *57*, 1086–1095. [CrossRef]
24. Dellacorte, C. Stiffness and Damping Coefficient Estimation of Compliant Surface Gas Bearings for Oil-Free Turbomachinery. *Tribol. Trans.* **2011**, *54*, 674–684. [CrossRef]
25. Howard, S.A. Rotordynamics and Design Methods of an Oil-Free Turbocharger. *Tribol Trans.* **1999**, *42*, 1. [CrossRef]
26. Walton, J.F.; Tomaszewski, M.J.; Heshmat, H.; ASME. *The Role of High Performance Foil Bearings in Advanced, Oil-Free, High-Speed Motor Driven Compressors*; Rochester: New York, NY, USA, 2003; pp. 411–417.
27. Sun, Y.H.; Cheng, W.J.; Chen, Z.F.; Zhuo, M.; Yu, L. Centrifugal Air Compressor Cooling Structure Driven by Ultra-High Speed Permanent Magnet Motor. C.N. 201310181042.5, 15 April 2015.
28. McAuliffe, C.; Dziorny, P.J. Bearing Cooling Arrangement for Air Cycle machine. U.S. Patent 5113670A, 19 May 1992.

29. Walton, J.F., II; Tomaszewski, M.J.; Heshmat, C.A. On the Development of an Oil-Free Electric Turbocharger for Fuel Cells. In Proceedings of the 2006 ASME Turbo Expo: Power for Land, Sea & Air, Barcelona, Spain, 1 January 2006.

30. Sun, Y.G.; Yu, X.M.; Liu, H.T. Analysis of Dynamic Stiffness and Damping of Partial Porous Aerostatic Thrust Bearings. *Appl. Mech. Mater.* **2009**, *875*, 596–600. [CrossRef]

31. Su, B.; Hou, Y.; Zhao, H.L.; Zhou, Q.; Chen, C.Z. Feasibility analysis on self-acting gas bearings of air condition system of plane. *Cryogeni* **2008**, *3*, 50–54. (In Chinese)

32. Feng, K.; Zhao, X.; Guo, Z. Design and structural performance measurements of a novel multi-cantilever foil bearing. *Proc. Inst. Mech. Eng. Part C J. Mech. Eng. Sci.* **2015**, *229*, 1830–1838. [CrossRef]

33. Feng, K.; Hu, J.; Liu, W.; Zhao, X.Y. Structural Characterization of a Novel Gas Foil Bearing with Nested Compression Springs: Analytical Modeling and Experimental Measurement. *J. Eng. Gas Turbines* **2016**, *138*. [CrossRef]

34. Li, C.L.; Du, J.J.; Yao, Y.X. Study of load carrying mechanism of a novel three-pad gas foil bearing with multiple sliding beams. *Mech. Syst. Signal Process.* **2020**, *135*. [CrossRef]

35. Luo, Y.X.; Zhang, J.Y. Analysis of gas film thermal characteristics of gas thrust bearing. *Mach. Build. Autom.* **2019**, *48*, 39–42. (In Chinese)

36. Liu, S.X.; Ma, X.Z. Dynamic Properties Analysis of Compliant Foil Aerodynamic Bearings Based on Spring Model. *Appl. Mech. Mater.* **2011**, *86*, 252–255. [CrossRef]

37. Du, K.Q.; Ma, X.Z. Research on the preparation and performances of MoS2-based lubricating coatings for gas bump foil bearings. *Lubr. Eng.* **2019**, *44*, 22–28. (In Chinese)

38. Hou, A.P.; Lin, P.C.; Wang, R.; L, J.X.; Hu, B. Performance of air-dynamic lubrication thrust bearing and experiment. *J. Aerosp. Power* **2018**, *33*, 1510–1518. (In Chinese)

39. Hou, Y.; Yang, S.J.; Chen, X.Y.; Chen, R.G.; Lai, T.W.; Chen, S.T. Application of compliant foil bearings in a cryogenic turboexpander. *Journal of Harbin Eng. Univ.* **2015**, *4*, 489–493. (In Chinese)

40. Xiong, L.Y.; Hou, Y.; Wu, G.; Wang, B.C.; Chen, C.Z. Analytical study of plate foil journal gas bearings. *J. Xi'an Jiaotong Univ.* **1998**, *9*, 56–59. (In Chinese)

41. Zhou, Q.; Hou, Y.; Cui, M.X.; Chen, C.Z. Analysis of new aerodynamic complaint foil thrust gas bearing. *J. Xi'an Jiaotong Univ.* **2006**, *40*, 1032–1035. (In Chinese)

42. Hou, Y.; Zhu, Z.H.; Chen, C.Z. Comparative test on two kinds of new compliant foil bearing for small cryogenic turbo-expander. *Cryogenics* **2004**, *44*, 69–72. [CrossRef]

43. Cui, M.X.; Hou, Y.; Wang, L.Z.; Chen, C.Z. On the calculation of structural stiffness of compliant bump foil bearing. *Lubr. Eng.* **2006**, *5*, 57–59. (In Chinese)

44. Ma, B.; Sun, W.; Lai, T.W.; Zheng, Y.Q.; Chen, S.T.; Hou, Y. Experiment study on gas lubricated hydrodynamic bump type foil bearing. *J. Xi'an Jiaotong Univ.* **2014**, *1*, 118–122. (In Chinese)

45. Zhou, Q.; Hou, Y.; Chen, R.G. Experimental research on compliant foil thrust bearings for high speed turbine expander. *J. Xi'an Jiaotong Univ.* **2012**, *11*, 49–52. (In Chinese)

46. Lai, T.W.; Zheng, Y.Q.; Chen, S.T.; Hou, Y. Experimental study on the performance of multi-decked elastic support protuberant foil gas bearing. *J. Northeast. Univ.* **2014**, *35*, 415–418. (In Chinese)

47. Lai, T.; Guo, Y.; Wang, W.; Hou, Y. Development and Application of Integrated Aerodynamic Protuberant Foil Journal and Thrust Bearing in Turboexpander. *Int. J. Rotating Mach.* **2017**, *2017*, 1–12. [CrossRef]

48. Guo, Y.; Lai, T.W.; Zhao, Q.; Chu, Y.; Chen, S.T.; Hou, Y. Experimental study of multi-leaf foil bearing inhigh-speed turbo machinery. *J. Xi'an Jiaotong Univ.* **2020**, *54*, 40–45. (In Chinese)

49. Guo, Y.; Hou, Y.; Zhao, Q.; Ren, X.H.; Lai, T.W. Application of multi-leaf foil bearings in high-speed turbo machinery. *J. Adv. Mech. Des. Syst. Manuf.* **2020**, *14*. [CrossRef]

50. Sheng, C.C.; Yang, Y.; Xie, H.T.; Gao, W.H.; Luo, G.Q.; Chen, S.T.; Hou, Y. Development and application of aerodynamic foil bearings in high-speed turbo machinery of airborne ECS. *Lubr. Eng.* **2020**, *45*, 86–90. (In Chinese)

51. Lv, P. Design and Experiment Study on Air Cycle Machine Rotor System Supported by Gas Foil Bearing. Master's Thesis, Hunan University, Changsha, China, 2017. (In Chinese).

52. Deng, W.M. Application of Air Bearing on Turbine Cooling Unit and its improvement. *Equip. Manuf. Technol.* **2014**, *4*, 140–142. (In Chinese)

53. Shu, X.J.; Xu, G.; Zheng, Y.Q.; Cui, H.L.; Lan, H. Key Technologies for centrifugal blowers with foil bearings and high speed magnet motor. *Compressor Blower Fan Technol.* **2017**, *59*, 36–42. (In Chinese)

54. Xie, Y.Q. Design and Experimental Study of an Oil Free Turbocharger. Master's Thesis, Hunan University, Changsha, China, 2015. (In Chinese)

55. Liu, W.H.; Lv, P.; Yu, R.; Feng, K. Design and experimental research of the oil-free turbocharger. *J. Mech. Eng.* **2018**, *54*, 129–136. (In Chinese) [CrossRef]

56. Lin, H.; Geng, H.; Yu, L.; Zhou, J.; Li, H. Electromechanical Analysis and Experiment for a High-speed Permanent Magnet Synchronous Compressor Supported by Elastic Foil Bearing. In Proceedings of the 2019 IEEE/ASME International Conference on Advanced Intelligent Mechatronics (AIM), Hong Kong, China, 8–12 July 2019.

57. Geng, H.P.; Cheng, W.J.; Yu, L. Cooling Structure of Motor-Driven High-Speed Centrifugal Air Compressor. CN:201110258998: A, 18 January 2012. (In Chinese)

58. Deng, Z.H. Design and Experimental Study of an Oil-Free High Speed Compressor for Fuel Cell. Master's Thesis, Hunan University, Changsha, China, 2016. (In Chinese)
59. Jin, C.Y.; Xu, F.C.; Wu, M.L.; Fan, J.Y.; Xin, J. Calculation and Analysis of Air Bearing Centrifugal Compressor Rotor Dynamics for Fuel Cells. *Automob. Parts* **2020**, *11*, 6–9. (In Chinese)
60. Xu, F.C.; Sun, Y.; Zhang, G.H.; Liu, Z.S. Effect of bump structural friction on the performance of bump foil bearing and rotor dynamic behavior: Experimental study. *Proc. Inst. Mech. Eng. Part J J. Eng. Tribol.* **2019**, *233*, 702–711. [CrossRef]

Article

Air-Based Contactless Wafer Precision Positioning System

Rico Hooijschuur, Niranjan Saikumar, S. Hassan HosseinNia and Ron A. J. van Ostayen *

Department of Precision and Microsystems Engineering, Delft University of Technology, Mekelweg 5, 2628 CD Delft, The Netherlands; ricohooijschuur@hotmail.com (R.H.); niranjans87@gmail.com (N.S.); S.H.HosseinNiaKani@tudelft.nl (S.H.H.)
* Correspondence: R.A.J.vanOstayen@tudelft.nl; Tel.: +31-15-2781647

Featured Application: In this paper an improved contactless handling and positioning method for thin substrates is proposed based on air bearing technology. Being fully contactless, the potential application of this work can be found in the high-tech industry where thin, flexible, but at the same time fragile substrates need to be handled with minimal risk of contamination, damage, or even breakage.

Abstract: This paper presents the development of a contactless sensing system and the dynamic evaluation of an air-bearing-based precision wafer positioning system. The contactless positioning stage is a response to the trend seen in the high-tech industry, where the substrates are becoming thinner and larger to reduce the cost and increase the yield. Using contactless handling it is possible to avoid damage and contamination. The system works by floating the substrate on a thin film of air. A viscous traction force is created on the substrate by steering the airflow. A cascaded control design structure has been implemented on the contactless positioning system, where the inner loop controller (ILC) controls the actuator which steers the airflow and the outer loop controller (OLC) controls the position of the substrate by controlling the reference of the ILC. The dynamics of the ILC are evaluated and optimized for the performance of the positioning of the substrate. The vibration disturbances are also handled by the ILC. The bandwidth of the system has been improved to 300 Hz. For the OLC, a linear charge-coupled device has been implemented as a contactless sensor. The performance of the sensing system has been analysed. During control in steady-state, this resulted in a position error of the substrate of 12.9 μm RMS, which is a little more than two times the resolution. The bandwidth of the OLC is approaching 10 Hz.

Keywords: air-bearing; precision positioning; tribotronics

Citation: Hooijschuur, R.; Saikumar, N.; HosseinNia, S.H.; van Ostayen, R.A.J. Air-Based Contactless Wafer Precision Positioning System. *Appl. Sci.* **2021**, *11*, 7588. https://doi.org/10.3390/app11167588

Academic Editors: Terenziano Raparelli, Federico Colombo, Andrea Trivella and Luigi Lentini

Received: 25 July 2021
Accepted: 14 August 2021
Published: 18 August 2021

Publisher's Note: MDPI stays neutral with regard to jurisdictional claims in published maps and institutional affiliations.

1. Introduction

In the high-tech industry products such as displays, solar panels, and chips are produced. These products are made out of brittle materials where thickness is an important factor. The thickness of the raw material is a factor in the total costs. Additionally, the efficiency of the production process can be increased by increasing the size of the substrates. This is why there was a trend where the substrates increased in size while decreasing in thickness [1]. However, this trend came to a halt due to the percentage of damaged and broken wafers during the manufacturing process.

A possible solution is to handle the products without mechanical contact, floating the substrate on a layer of air. This type of air-based levitation is known from air-bearing technology, where pressurized air forms a thin air film that separates the two surfaces. This concept is an alternative to mechanical contact, where friction, stick-slip, and wear have to be minimized. Furthermore, the layer of air can support the whole surface of the substrate. This is an improvement over the point contact that is currently being used [2], since an air bearing limits particle generation and resulting contamination and forces on the substrate. This will increase the yield from a single wafer.

Research at the Delft University of Technology has generated multiple concepts where substrates are controlled on a controlled layer of air. The concepts vary in the way the air is controlled, i.e., by variation of the pressure inlet [3], deformation of the surface to control the airflow [4], or variation of the outlet restriction [5]. This technology has also been used to levitate and transport different kinds of materials [6,7], because the air-based levitation is not limited to specific substrate material properties.

When the substrate is floating in a positioning system it is necessary to measure and control its position. In the research cited above, the substrates were modified to measure their position. For example, gratings were added to the wafers. Research has been done on contactless positioning [6,8], but sensing systems that do not require any modifications to the substrate have been limited in either bandwidth or resolution.

This paper presents an air-bearing positioning system where no modifications are made to the substrate, in line with how such an air-bearing motion system is to be used in the high-tech industry. The implementation of a charge-coupled device (CCD) is proposed for noncontact position measurement. The sensing concept is implemented in the contactless actuation system and the results are presented.

In Section 2, the state of the art is presented. Section 3 continues with a focused and detailed description of the existing contactless actuation concept based on deformable surfaces as is relevant for the rest of the paper. The dynamics response of the actuator is evaluated and a controller is designed in Section 4. Section 5 shows the proposed sensor concept with performance analysis. The sensor is implemented in the contactless actuation system and the results are presented in Section 6.

2. State of the Art

A survey of recent developments in the field of contactless manipulation is presented in [9]. The use of airflow for manipulation and control of the localised deformation of wafers has also been studied in [10], while a pick-up head for wafers has been designed using similar concepts in [11]. Ultra-sound air film technology has also been studied for this application in [12]. However, while this technology provides a viable alternative for contactless pick-up of wafers, it is mainly foreseen to be used in a pick-up head and not for complete manipulation and/or transportation of wafers using airflow.

The start of the development of the new contactless actuation concept at TU Delft was in 2007 [3]. The concepts use the same physics to control the position of the substrate. First, the substrate is levitated on a layer of air. Subsequently, the substrate is preloaded in the out of plane direction with a controlled vacuum source. This increases the out of plane stiffness at the fly height. This height can be adjusted by changing the force the vacuum and pressure source are applying.

The in-plane actuation is achieved by creating an airflow tangential to the substrate surface. The air is flowing from the high-pressure area to the lower pressure following the path of the least resistance. This flow of air exerts a viscous traction force on the substrate. This innovative idea has lead to multiple concepts that will be discussed in this section.

2.1. Variable Inlet Pressure

In 2007, Wesseling designed a contactless positioning system based on a fixed geometry and a variable inlet pressure to control the thin film airflow in his PhD thesis "Contactless Positioning Using an Active Air-Film" [3]. Figure 1a shows a cross-section of one of the actuator cells, where a substrate is floating on top of the pressurized air.

The air will flow from the inlet $p+$ with high pressure to the outlet $p-$, where the air is removed from the cell at a sub-ambient pressure. The viscous shear of the airflow creates a traction force on the substrate. The direction of the flow within a cell is controlled by changing the pressure at the different inlets. Four actuator cells have to work together to actuate the substrate in three degrees of freedom. The flow of the different cells is separated by dams. Four actuator cells are shown in Figure 1b.

The final system was used to position a 4-inch silicon wafer, a demonstrator as shown in Figure 1c. A bandwidth of 50 Hz was realized using a PID controller. The system had a maximum acceleration of 600 mm/s^2, with an error of 6 nm (1σ), while using an external active vibration isolation system. The concept's performance was limited by the manufacturing tolerances of the system, since this placed a lower limit on the fly height of 15 µm. A lower fly height would result in a higher actuation force. The other limiting factor was the time delay caused by the pressure lines between the control valve and the pressure inlet in the cell.

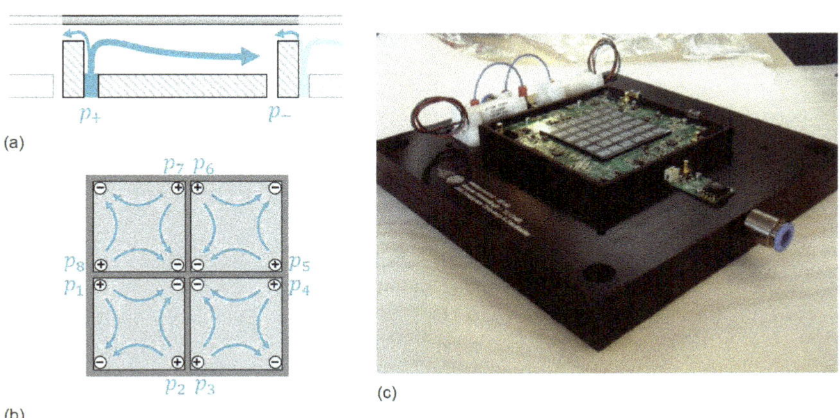

Figure 1. Working principle of the first generation. (**a**) Shows the cross-section of an actuator cell. (**b**) Top view. (**c**) The pressure variation demonstrator. For reference, the size of 6 × 6 actuator array in the centre of the image is 60 mm × 60 mm [3,13].

2.2. Variable Outlet Restriction

In 2017, Verbruggen designed the third generation of the contactless actuator which uses an outlet restriction to create the traction force. The dams can be actuated to restrict the flow at one side, while the flow is opened up at the other side. The concept is similar to the variable inlet pressure design. The advantage of this system is that the restriction is close to the chamber. This is an improvement, because there is no delay present at the supply lines, which limited the variable inlet pressure stage.

A demonstrator was built to prove the concept. Seven circular pockets have been produced. The substrate was controlled in open-loop. Using feedforward average angular accelerations have been measured of 11 rad/s^2.

2.3. Transportation

Besides positioning a substrate on a stage, this technology can also be used for transportation thus further reducing the need for mechanical contact in a typical production system. In 2016, Vagher investigated the construction of a passive contactless conveyor [6]. In this research, a cost-effective manufacturing technique for the passive conveyor surface was explored.

Vagher's conveyor concept moves substrates in 2D and measured the position of the substrate using an automated vision system. The vision system resulted in a resolution of 30 µm at 60 Hz. In continuation, Snieder investigated the transition between active and passive surfaces, which makes the transportation concept more versatile [7].

2.4. Deformable Surface

In 2011, Vuong explored a new concept regarding contactless positioning, "Air-based contactless actuation system for thin substrates" [4]. The concept is based on multiple tilted surfaces to steer the airflow, and therefore the force on the substrate. The advantage of the tilting surface is that a change in the geometry directly generates a change of the shear force on the substrate. This concept does not suffer from the delay between the controlling action and force as seen in the first generation. In this concept, the in- and outlet pressure can be kept constant. The concept where the surfaces are tilted by bending the shafts is shown in Figure 2. This concept is used in this paper for the precision contactless positioning of the wafer and is explained in detail in the next section.

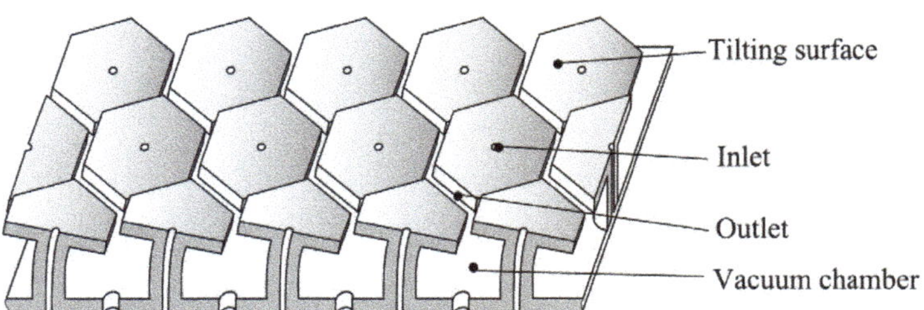

Figure 2. The tilting surface concept, the surfaces are titled by bending the shafts [4].

3. Flowerbed

A demonstrator was built based on the bendable shaft concept. However, instead of bendable shafts, two membranes are used which allows the shafts to tilt. This reduces the force requirement of the actuator. The concept is shown in Figure 3. The top membrane connects the flowers to the world. The bottom membrane is attached to the movable plate, which is connected to an actuator. The top membrane also functions as a seal for the upper chamber, which is connected to the vacuum pressure outlet. The inlet pressure comes in through the hollow centres of the shafts.

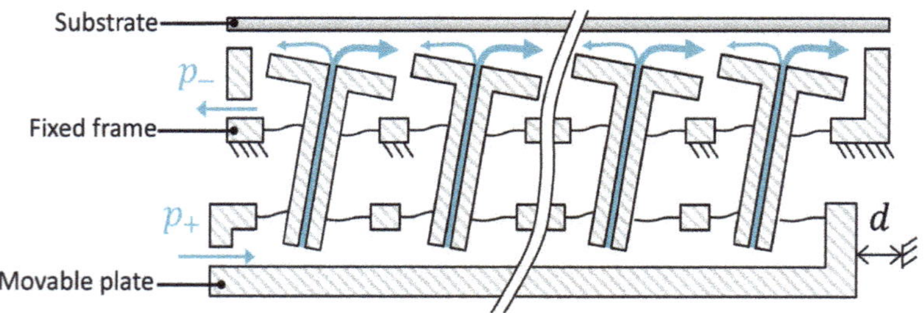

Figure 3. Schematic of the second generation contactless actuator, showing the flow of air through the system, for a displacement d of the movable plate [13].

Figure 4 shows a cut-through view of the Flowerbed. The Flowerbed has 61 hexagonal heads, with an in-radius of 14 mm. It is nicknamed the Flowerbed because the surface looks like flowers that tilt towards the sun. The hexagon shape of the flowers was selected because it provides a full packing of the surface in combination with an optimal circumference to surface area ratio, and thus a maximum force for a given airflow. Furthermore, the figure also shows the fixed plate and movable plate, where both consist of two plates that are clamped together with a 50 µm thick membrane in between. The membrane is also clamped in the flowers. To better align the flowers, the movable plate is preloaded downwards with three springs, with a force of approximately 100 N. After assembly, the final step was the spark erosion of the surface with wire-cut EDM technology to further level the surface.

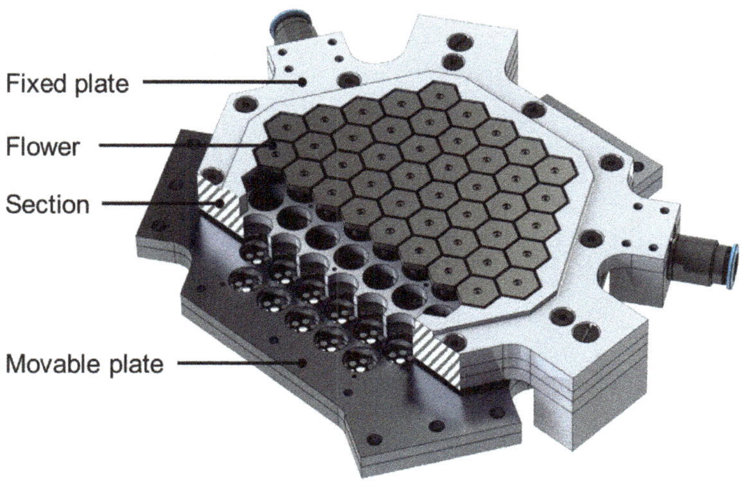

Figure 4. A cross-section of the Flowerbed showing the fixed plate, flowers, and movable plate. Source: [13].

Figure 5 shows a single flower in both a neutral and actuated position. The thin film airflow is visualized in the area between the effective surface of the flower and the substrate. The combination of the positive and negative pressures provides stiffness in the out-of-plane direction at the fly-height h.

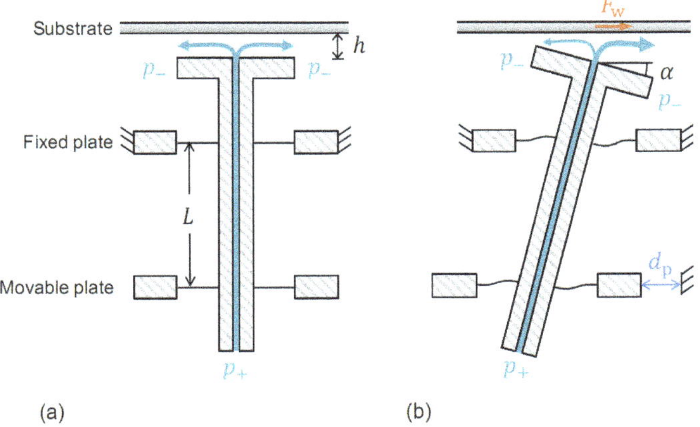

Figure 5. Schematic of a single flower of the Flowerbed in (**a**) neutral position and (**b**) actuated position. The force on the wafer is visualized in the actuated position [13].

The flowers are tilted with an angle α, through a displacement of the movable plate, as shown in Equation (1). Where d_p is the displacement of the movable plate and L is the distance between the movable plate and the fixed plate.

$$sin(\alpha) = \frac{d_p}{L} \approx \alpha \tag{1}$$

when tilted the majority of the airflow will follow the path of least resistance. This flow will, in turn, exert a viscous traction force F_w on the wafer in addition to the out of plane stiffness.

The tilting angle of the flower has been measured to be proportional to the displacement of the movable plate up to 1000 Hz, without delay. The tilt response between the individual flowers is similar, there is only a small deviation in the gain. The values of gain vary between 71 rad/m and 80 rad/m, where the predicted value is 77 rad/m [4]. Furthermore, Vuong experimentally validated that the pneumatics can be modelled by a proportional gain, up to at least 400 Hz. This means that if the movable plate position is known, the force exerted on the wafer is known.

The stiffness of the Flowerbed is different for the translation and rotation of the movable plate. Each flower has an elastic hinge at the top and bottom, which provide stiffness against rotation. Figure 6, shows the behaviour of the Flowerbed during the translation and rotation of the movable plate. The stiffness and inertia values for translation and rotation result in two suspension eigenvalues. These values have been calculated and experimentally validated by Jansen [13]. The important properties are summarized in Table 1.

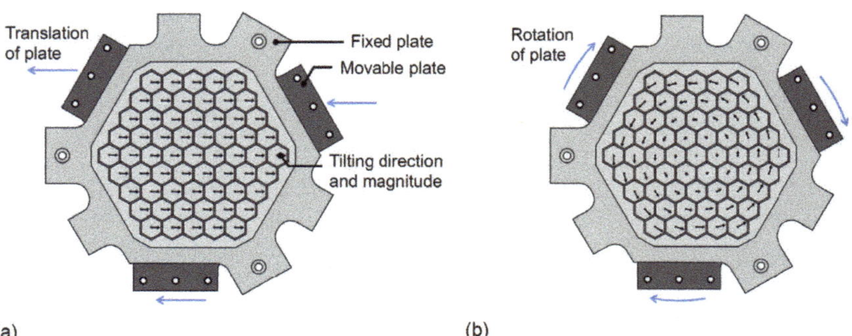

Figure 6. Schematic of the flowerbed during (**a**) translation and (**b**) totation [13].

Table 1. Flowerbed properties.

Property		Value
Translational stiffness	k_{eq}	200 kN/m
Rotational stiffness	k_θ	2.1 kNm/rad
Equivalent mass	m_{eq}	1.9 kg
Equivalent inertia	I_{eq}	5.2×10^{-3} kg m^2
Translational eigenmode	$f_{0,d}$	51.6 Hz
Rotational eigenmode	$f_{0,\theta}$	101 Hz

3.1. Flowerbed Actuation

In 2016, Krijnen designed a reluctance actuator to actuate the movable plate of the Flowerbed [14,15]. Reluctance actuators can provide a high force, but they also have a high mass. With the use of fractional order control, this resulted in a control bandwidth for the Flowerbed of 100 Hz. The bandwidth was limited by the eigenmodes caused by the actuator as shown in Figure 7. Figure 8 shows the flapping mode of the actuators that was limiting the bandwidth. The position of the wafer is measured using optical sensors. This required that gratings are attached to the wafer. The position of the wafer is controlled with a bandwidth of 60 Hz.

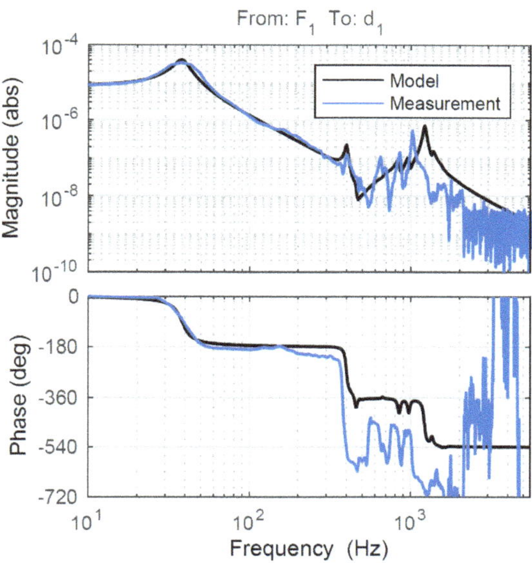

Figure 7. The simulated and measured frequency response function of the Flowerbed with the reluctance actuator [14].

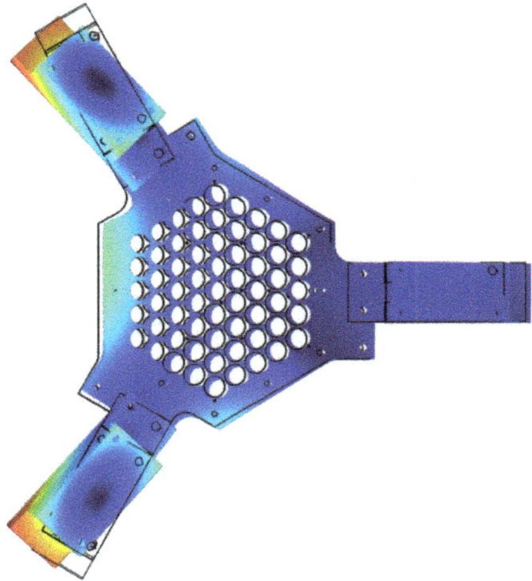

Figure 8. Flapping modeshape of the actuators at 1080 Hz (measured) [14].

To increase the bandwidth of the system, Jansen redesigned the actuator in 2018. The new actuator was designed with special care for dynamic performance. Instead of traditional hinges, a design with notch flexures is chosen. The flexures provide a high stiffness in the out of the plane direction and work without backlash. The reluctance actuators have been replaced with Lorentz actuators, because they have a linear force current relation. Furthermore, the permanent magnets were placed on the stationary frame, and the coils on the mover, resulting in a minimal moving mass.

Figure 9 shows the chosen guidance for the movable plate. In the figure, the application point of the force and sensor are shown, as well as the notch flexure and guide linkages. The sensors that measure the actuator displacement are placed close to the actuator.

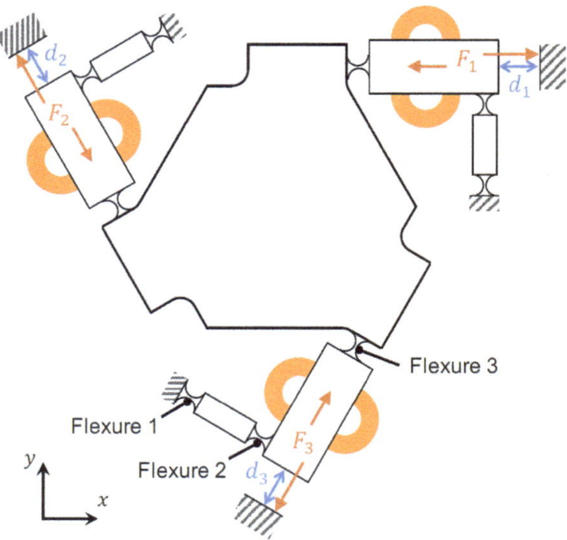

Figure 9. Compliant mechanism design used to house the Lorentz actuators. Adapted from [13].

Figure 10 shows the Flowerbed including the realized actuator. From top to bottom it shows the connection between the actuator and movable plate. The coil is used to generate the force on the movable plate. The sensor is placed directly behind the actuator. The actuators have a relatively small motion range thus the wires running from the coil to the base plane are no problem. Furthermore, the stiffness is small compared to the stiffness of the movable plate [13].

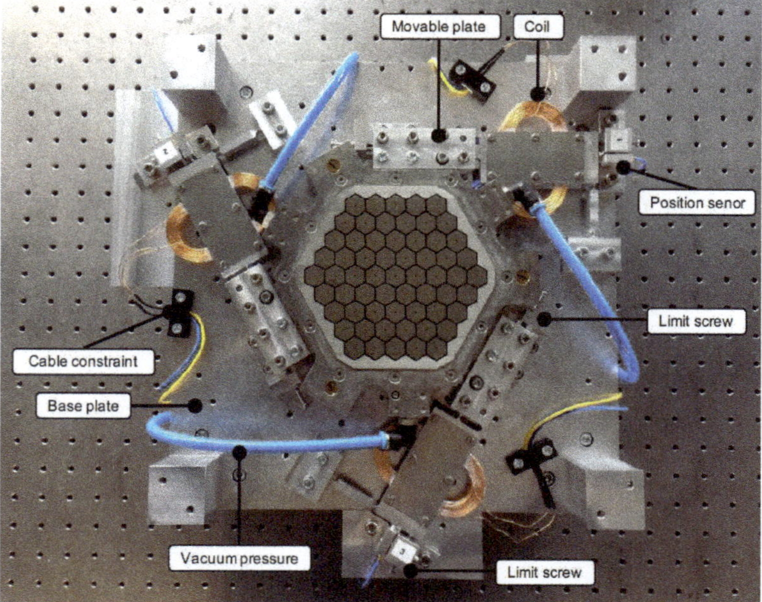

Figure 10. Photograph of the Flowerbed with the compliant mechanism and Lorentz actuator design [13].

The position of the actuators is measured using Micro-Epsilon CapaNCDT CS08 capacitive sensors, which have a range of 800 μm. With the 16 bits ADC this resulted in a resolution of 24.4 nm. The noise is measured at 0.6 μm peak-to-peak, with an RMS value of 72 nm.

4. Flowerbed Dynamics and Control

In the previous sections, the state of the art has been presented. The bandwidth of the compliant actuator design was placed on 233 Hz. No sensor was implemented to measure the position of the wafer. In this section, the dynamic response of the system is presented. A controller is designed and the closed-loop results are discussed. At last, the possibilities for improvement are discussed.

4.1. Flowerbed Dynamics

Figures 11 and 12 show the results from the identification where all the straight transfers and cross transfers are plotted. The directions of force and position sensor are inverted and account for the 180° phase seen. Both transfers show a quick drop in phase which is the limiting factor for the bandwidth. Jansen suggested this could be the effect of the capacitive sensing system [13]. This large phase drop was investigated as part of this work. After investigation of the delay sources in the system— amplifier, capacitive sensors, sampling time, and coil dynamics—the main sources of the delay have been identified. The phase lag is due to a combination of the 20 kHz sampling which produces a phase lag and the magnetic field which is lagging the current. The lagging field is due to the eddy currents in the mover which will create a magnetic field opposing the movement. When the magnetic field of the coil and eddy current are summed the field lags that of the coil alone. This provides a time delay seen in the bode diagram as a lagging phase. Furthermore, the eddy current damping provides damping of magnitude proportional to the velocity of the mover.

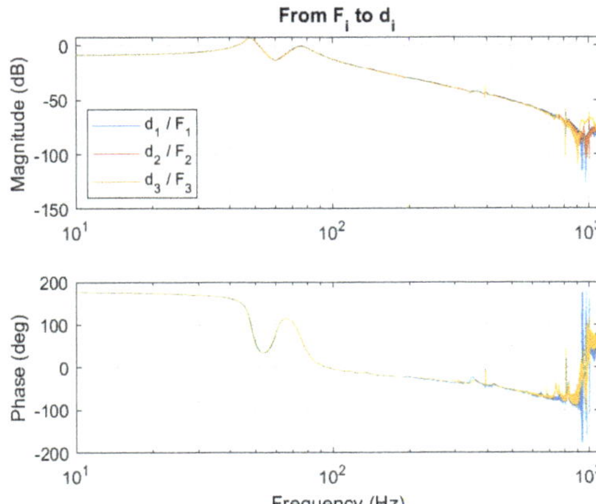

Figure 11. Frequency response function from actuator force to actuator displacement, the phase is shifted upwards by 180 degrees since the position is measured in opposite direction to applied force.

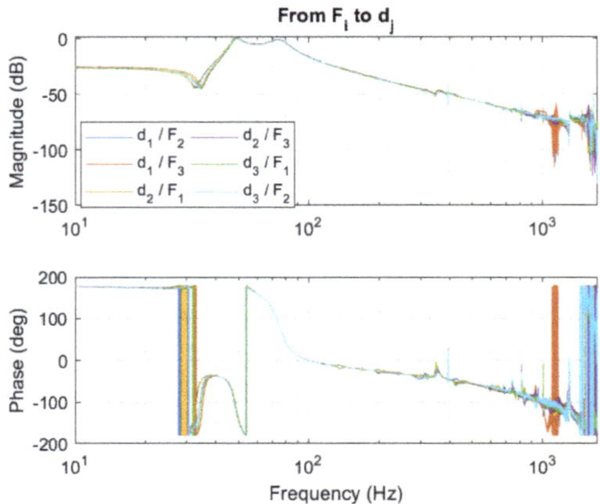

Figure 12. The cross-coupled frequency response functions of the Flowerbed, the phase is shifted upwards by 180 degrees since the position is measured in opposite direction to applied force.

Figure 13 shows the mover in the magnetic field of the permanent magnets. The mover consists of an aluminium plate with spaces cut out for the coils and a top plate to hold them in place. A solution to limit the intensity of the generated eddy currents is to laminate the mover. Another method is to look at the possibility of flux feedback for the amplifier of the system, to compensate for the effects of the eddy current [16]. These solutions have not been implemented in this work and are left for future work.

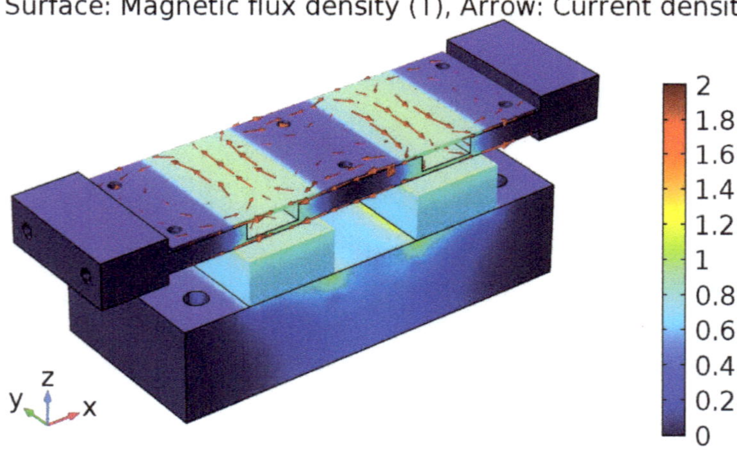

Figure 13. Simulation showing the magnetic flux density (in T on the colorbar) and the eddy current density (arrows) [13].

4.2. Flowerbed Control

The controller of the inner loop will control the position of the actuator to the corresponding sensor. The measured position will be controlled using a feedforward controller and feedback controller. The diagram is shown in Figure 14.

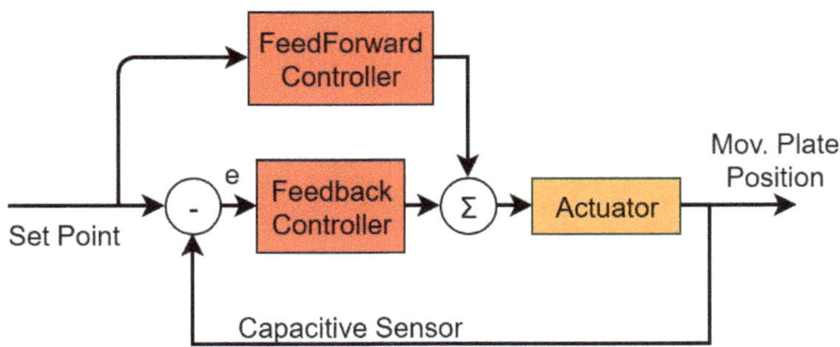

Figure 14. Movable plate control strategy (ILC).

For the feedback controller, a PID controller is designed for a bandwidth of 300 Hz. The parameters are shown in Table 2, the gain for actuator 2 is different to compensate for the slightly different stiffness at the bandwidth. The controller creates a phase margin of 27°, gain margin of 32.7 dB, and a modulus margin of 11.3 dB.

Table 2. ILC-PID parameters.

Parameter		Actuator 1	Actuator 2	Actuator 3
K	N/m	15.2	14.8	15.2
f_i	Hz	10	10	10
f_d	Hz	50	50	50
f_t	Hz	2000	2000	2000
f_f	Hz	2500	2500	2500

The feedforward controller has been designed using a simplification of the plant. The feedforward controller is the inverse of this plant made proper with two second-order low pass filters at the frequency of the bandwidth. The closed-loop response of the system with the PID controller and feedforward is shown in Figures 15 and 16. The feedforward controller is given in Equation (2).

$$FF = \frac{\left(\frac{s}{2\pi 300} + 1\right)\left(\frac{s^2}{(2\pi 49)^2} + \frac{0.1\,s}{2\pi 49} + 1\right)}{0.37\left(\frac{s^2}{2\pi 300^2} + \frac{2\,s}{2\pi 300} + 1\right)^2} \cdot \frac{\left(\frac{s^2}{(2\pi 75)^2} + \frac{0.17\,s}{2\pi 75} + 1\right)}{\left(\frac{s^2}{(2\pi 60)^2} + \frac{2\,s}{2\pi 60} + 1\right)} \tag{2}$$

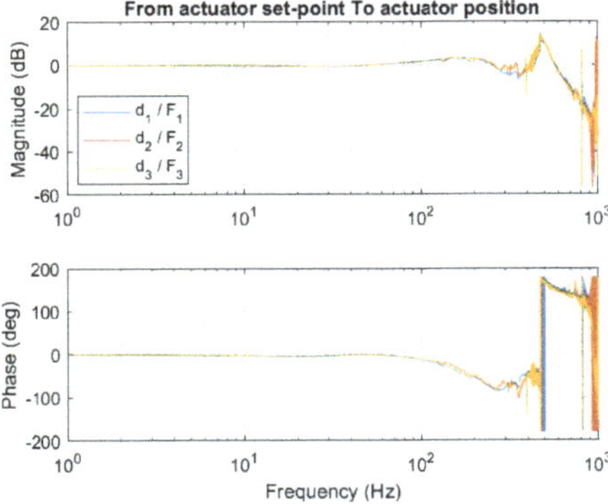

Figure 15. Closed loop frequency response of the actuators, from actuator reference position to measured position.

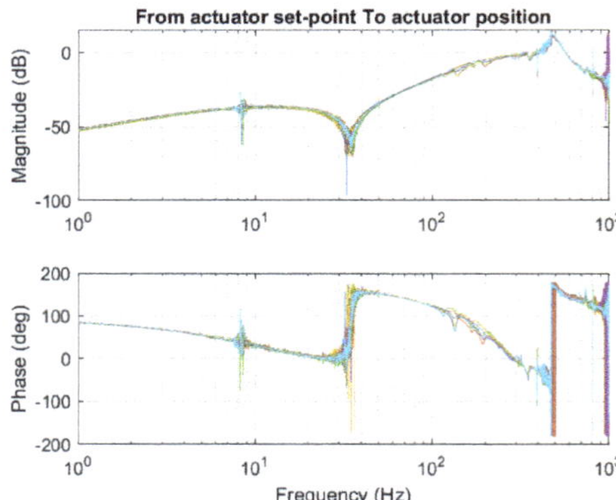

Figure 16. Closed loop cross coupled frequency response of the actuators, from actuator reference position to measured position.

The closed-loop response of the system is similar for all three actuators. The response provides a good basis for the outer loop controller up to, at least 100 Hz. The system shows strong coupling at 470 Hz, this could be improved using a decoupling filter or by prefiltering the input. Because the bandwidth of the outer loop is limited, it is chosen to use the outer loop controller to prefilter the reference signal.

5. Wafer Sensing

The wafer floating on air is a floating mass with a viscous damping system, thus to gain control over the position of the wafer it is necessary to have a collocated sensor. For this sensor in the outer loop, there are some restrictions, that influence the sensor choice. For the high-tech industry, it is unappealing to use a wafer modified to accommodate the sensing solution. Thus, it is preferred that the wafer is not altered.

As discussed in the state of the art, different sensor techniques have been used in the past, such as techniques based on capacitive sensors, edge detection using charge-coupled devices (CCD), and vision and optical encoders. The best speed and resolution was achieved using optical encoders with gratings on the wafer. However, this is an alteration to the wafer. Without altering the wafer edge detection using the linear CCD sensor array is the best choice. Edge detection can reach sampling speeds up to multiple kHz while having a sub-micrometre resolution that can be achieved using sub-pixel interpolation. This will be discussed later in this section.

5.1. Linear Sensor Array

A linear CCD has been selected to measure the edge position of the wafer. It consists of an array of pixels that measure the intensity of the light. The change from light to dark is used to find the edge of the wafer. The concept is shown in Figure 17. As discussed in the state of the art, Wesselingh had implemented CCD sensors but used a somewhat different concept where noise from stray light limited the performance [3].

The selection of the CCD sensor is based on a trade-off between the resolution, sampling time, and range. The sampling time is limited by the available modules for the National Instruments CompactRio. The NI9201 is used which supports 500 kS/s. The TCD1103GFG has 1500 pixels of 5.5 μm by 64 μm on a 5.5 μm center. It supports a data rate of 2 MHz. With 1500 pixels and a low integration time this results in a sampling rate of almost 1.3 kHz, and with the limitation of the 500 kS/s this is lowered to 320 Hz. High power cool white LEDs are used as the light source. The intensity of the light is tuned to maximize the dynamic range.

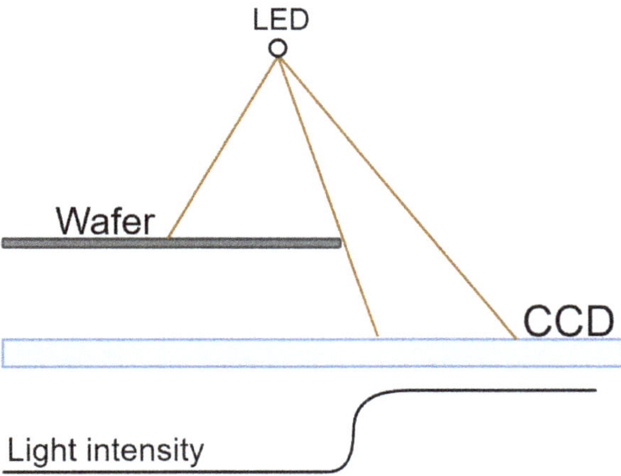

Figure 17. Linear CCD sensor concept to measure the edge of a wafer.

In this paper, the sensor output is compared to a threshold. This threshold is used to detect the transition from light to dark. Instead, a threshold a sub-pixel interpolation method could also be used. Using this method it is possible to increase the resolution of the sensors to better than 0.05 of the pixel size, while at the same time reducing the effects of noise [17–20]. Using the sub-pixel interpolation method the theoretic resolution of the chosen CCD could be as low as 0.11 µm. However, to limit the calculation time on the FPGA, it was chosen to implement the threshold model. Figure 18 shows the output of the CCD sensor. The threshold is placed at 0.6. When the threshold is reached the pixel number corresponds to the position on the CCD where the shadow ends.

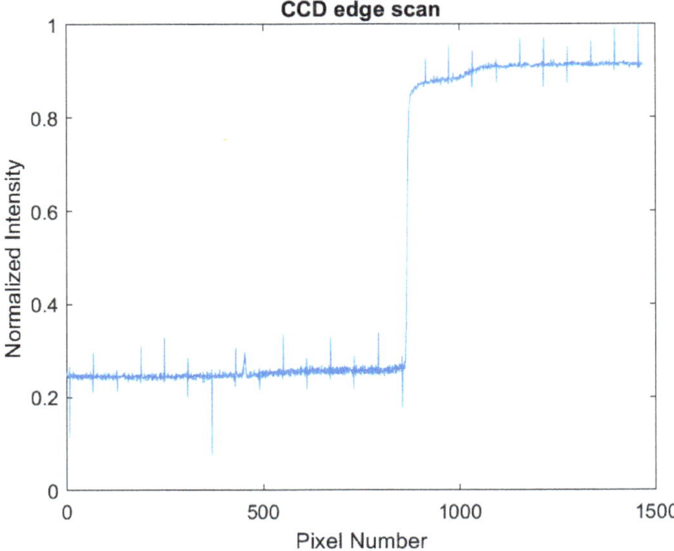

Figure 18. Output of the CCD sensor. It shows the edge of the wafer. The edge is a curve over 10 pixels.

When using a point light source, the measured position should be adjusted to get the real position. Figure 19 shows that the distance between the shadow and the wafer position is dependent on the position and height of the wafer. The wafer position is calculated using Equation (3a). Because the height of the wafer is not fixed a disturbance could enter the system through a variation in the height. The size of the disturbance can be estimated by Equation (3a).

The difference in fly-height is estimated to be in the order of micrometers. When minimizing the height between the wafer and CCD the disturbance is estimated to be within 0.1 of the pixel size for large movements, which is acceptable when using the threshold model, if sub-pixel interpolation is used it is better to remove the height by using a collimated light source or by adding another LED. With two LEDs, another data point is known which can be used to eliminate the height from the equation as shown in Equation (3b) [21].

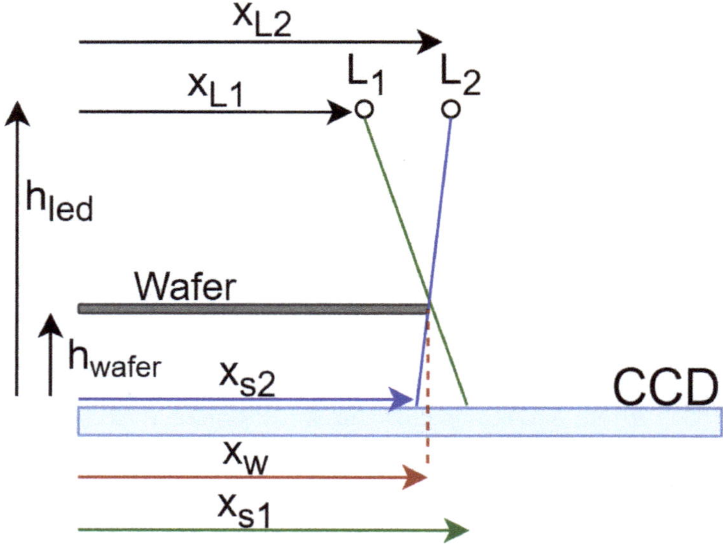

Figure 19. A CCD sensor with 2 LEDs, showing the actual position of the wafer x_w, the shadow created by the first led x_{s1} and second led x_{s2}.

$$X_w = \frac{h_{wafer}}{h_{led}}(x_{L1} - xs1) + s1 \tag{3a}$$

$$X_w = \frac{x_{L2}x_{s1} - x_{L1}x_{s2}}{-x_{L1} + x_{L2} + x_{s1} - x_{s2}} \tag{3b}$$

5.2. Performance Analysis

Before the implementation of the CCD sensor inside the contactless actuator, a performance analysis experiment was done. In these tests, the linearity of the sensor was tested, as well as the steady-state behaviour. The performance test was performed in an environment where disturbance sources from the external lighting, vibrations, and the electrical side were not removed. The steady-state response shows no drift and high precision, since the position returns to the same position. Figure 20 shows the performance analysis test where the position of the CCD is compared to a stage that has a resolution of 100 nm. The figure shows no drift and follows the position of the stage with an error of 3.0 μm RMS.

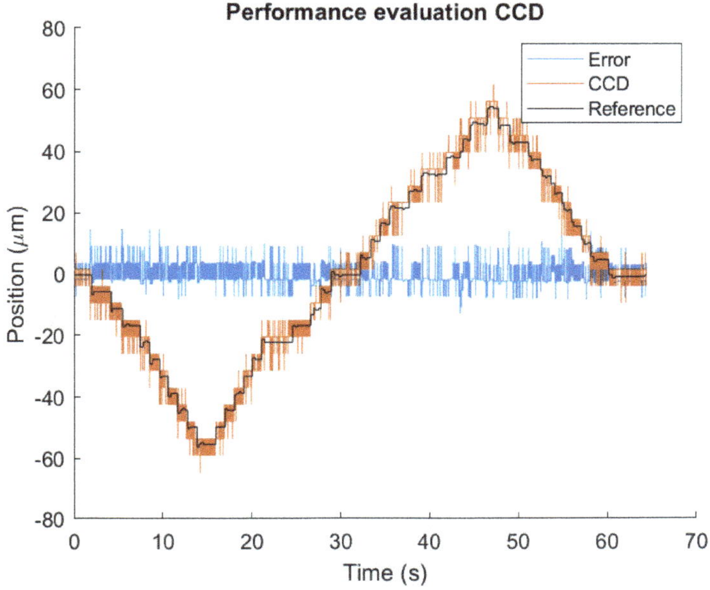

Figure 20. Performance of CCD sensor analysed for linearity. RMS error value 3.0 µm.

5.3. Metrology

The kinematics are used to convert the measured position from the sensors into the position of the wafer. Figure 21, shows the placement of the sensors. The kinematic equations are shown in Equation (4), where r is the radius of the wafer and 41.5 mm the distance between the sensors. The input of the kinematic equations is in millimetres and the output is in millimetres and radians. The photograph of the Flowerbed system with the integrated contactless sensing is shown in Figure 22.

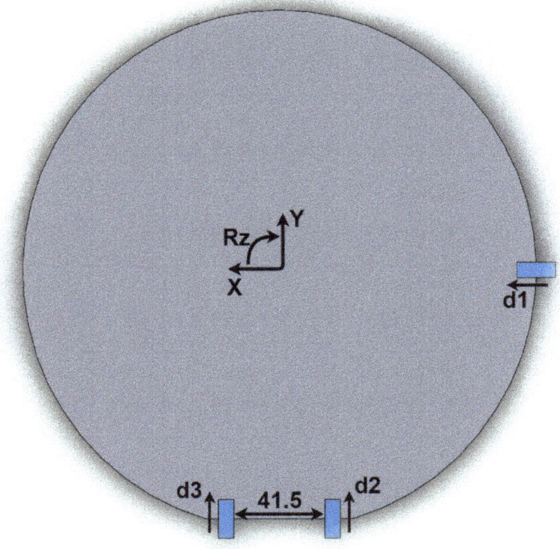

Figure 21. The sensor placement in reference to the wafer to be controlled.

$$y = \frac{d_2 + d_3}{2} \tag{4a}$$

$$x = d_1 - \left(r - \sqrt{r^2 - y^2} \right) \tag{4b}$$

$$Rz = \frac{d_3 - d_2}{41.5} \tag{4c}$$

The validation of the outer loop control was performed using a 200 mm dummy wafer. The dummy wafer was cut out of a 750 μm thick aluminium sheet according to the JEIDA standard. This means a circle with a radius of 100 mm with a flat of 57.5 mm. The aluminium wafer had a mass of 63 g, which is 8 percent more than a comparable 8 inch silicon wafer.

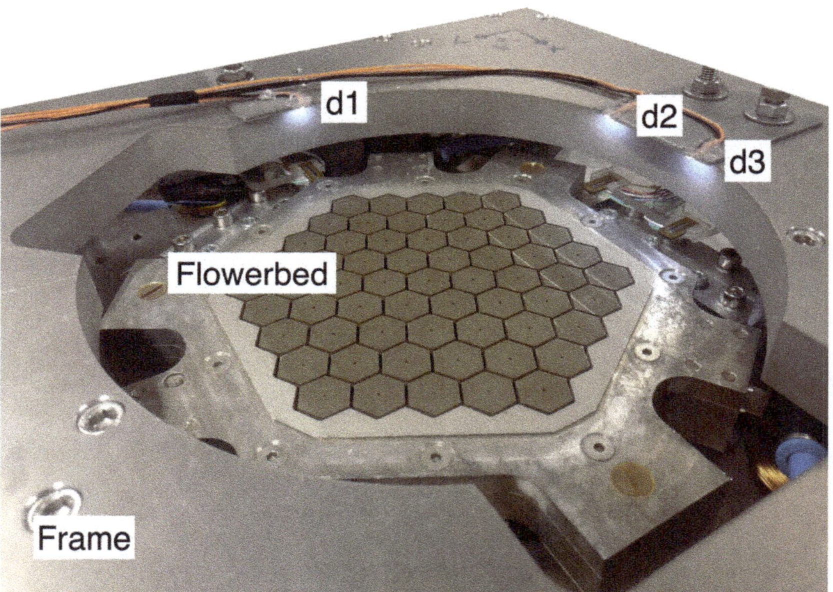

Figure 22. Photograph of the Flowerbed system with the implemented contactless sensing system without the wafer. The position of the sensors has been annotated according to Figure 21, and for reference, the diameter for the inscribed circle in each of the 61 hexagonal flowers is 14 mm.

6. Wafer Control

To control the position of the wafer, a second controller needs to be added to the inner loop Flowerbed control. This leads to a cascaded control structure as shown in Figure 23. The Flowerbed controller will be called the inner loop controller (ILC) from now on. The ILC will provide a stiff connection between the Flowerbed and actuator. The setpoint of the ILC will be controlled by the wafer controller, which is also called the outer loop controller (OLC). The kinematics and inverse kinematics are placed in the loop of the OLC, because the OLC has a lower bandwidth and thus more time for the calculations.

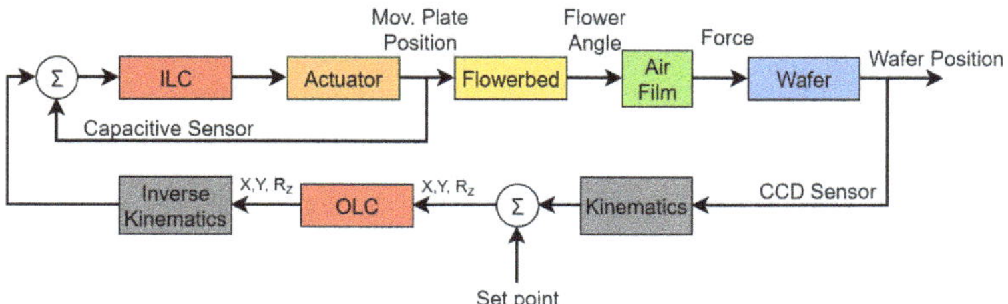

Figure 23. The control strategy for the flowerbed.

6.1. Pressure Control

For the operation of the Flowerbed, constant pressure and vacuum are crucial, even more so because of the non-uniform behaviour of the flowers on the surface for a given supply and vacuum pressure. As a result, setting a higher pressure will cause an extra force in the negative y-direction, whereas a lower vacuum will create an extra force in the positive y-direction.

To control the pressure, sensors and valves are placed in the supply lines. The dynamic behaviour of the pressure and vacuum supply has been identified around the operation pressure of 240 kPa supply and 6 kPa vacuum. The controller design is a proportional gain together with a second-order low pass filter with the values provided in Table 3. The lower bandwidth has been chosen because of the limitations of the solenoid valves, increasing the bandwidth further does not improve the performance. The closed-loop frequency responses for the pressure controls are shown in Figure 24. During control the steady-state error is measured, the error for the vacuum controller is 0.01 kPa (σ) and for the pressure controller 0.3 kPa (σ).

Table 3. Pressure controller properties.

Parameter	Pressure	Vacuum
Kp	5000	-1000
f_f	90	150

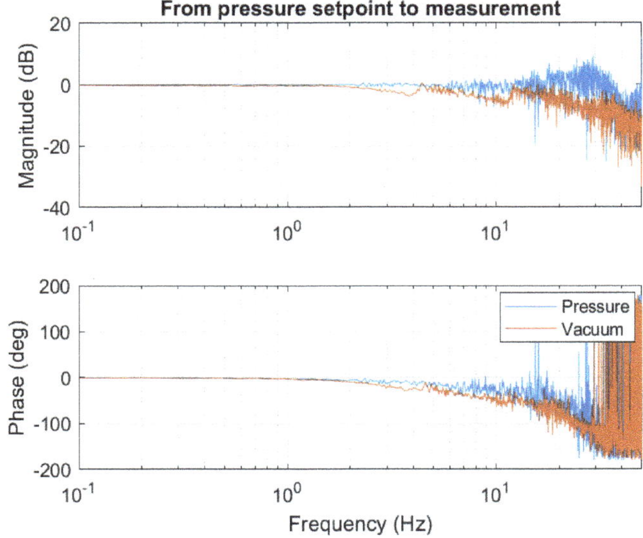

Figure 24. The closed loop frequency response of the Flowerbed pressure control.

6.2. Pneumatics

Previous experiments by Vuong showed that the air film dynamics can be described as a pure proportional gain up to at least 400 Hz, after which the measurement became unreliable [4]. The proportional gain of the system was determined at 1.4 mN/μm. In short, this means that if the position of the movable plate is known, the viscous traction force on the wafer can be calculated up to at least 400 Hz. This value of the proportional gain has been verified during the experiments of Krijnen [14]. Furthermore, Krijnen provided the displacement force graph for different supply pressures while maintaining the vacuum supply at ambient pressure minus 7 kPa.

Multiple experiments were done as part of this work with the values provided by Krijnen, however, the previously measured pneumatic gain could not be replicated again without mechanical contact. Multiple factors play a role here, the Flowerbed has been taken apart and reassembled multiple times to replace the actuators. Upon close inspection, it was noticed that multiple flowers were out of alignment. Furthermore, the supply pressure was leaking between the pressure sensor and the Flowerbed. This made the readings from the pressure sensor unreliable. The combination of these factors resulted in the fact that the noncontact criteria could not be guaranteed for a given pressure. Therefore, in this work, the pressure was manually tuned to obtain the best performance while maintaining the no-contact condition for a full movement of 60 μm for the movable plate. The optimum was found at a vacuum pressure at minus 6 kPa and a supply pressure of 245 kPa. As a result, the value of the pneumatic gain was less than optimal and was approximated at 0.3 mN/μm.

6.3. Outer Loop Controller

In the outer loop, two separate controllers have been designed, one for controlling the rotation at lower bandwidth and one for positioning the wafer in x and y at a higher bandwidth to maximize the performance of the total system. The parameters of the controller will be discussed later in this section.

The outer loop could not be identified using the open-loop response because the system consists of a floating mass with viscous damping. The identification of the system could be done by adding stiffness to the mass or through simulation. The simulated open-loop transfer function could be described by the transmissibility of the inner loop controller (T_{ILC}), the pneumatic gain of the system and the wafer, which is a mass system. This is shown in Equation (5), where C_{OLC} is the outer loop controller.

$$PC = \frac{T_{ILC}C_{OLC}}{m_{wafer}s^2} \tag{5}$$

For the controller, it is decided to use a PD controller with a low pass filter. The low pass filter must be placed before 300 Hz to cancel out the coupling in the inner loop. Because of the high gain of the mass-line, it was chosen to leave out the integrator.

$$K_p = \frac{m_{wafer}\omega_{bw}^2}{k_{pn}*4}; \tag{6}$$

The gain of the system was calculated using Equation (6). The factor 4 accounts for the frequency range of the derivative part of the controller as given in Table 4, which adds a gain of the same value. To start, the pneumatic gain k_{pn} values calculated by Krijnen were used. This gives an initial place to start after which the controller has been tuned iteratively. After the initial measurement, the new controller gain could be calculated to compensate for the lower pneumatic gain.

Table 4. Outer loop controller properties.

Parameter	Translation	Rotation
K_p	0.207	2.842
f_d	2.5	2.5
f_t	40	40
f_f	150	200

6.4. Performance

After tuning the controller for the inner loop, outer loop, and pressure, the performance of the positioning of the wafer has been measured. Figure 25, shows the closed-loop response of the system. The measurements for the cross-coupling between the x and y-axis are insignificant and thus not presented in the frequency domain. The error of the x and y-axis during tracking is presented in Figure 26. Furthermore, when tracking a sine wave with an amplitude of 0.33 mm at 0.5 Hz, the RMS error of x and y remained under 9 μm.

The error at steady-state is shown in Figure 27 with the RMS error values provided in Table 5. The periodic movement is caused by the correlation between the rotation and translational degrees of freedom. This is expected because the dummy wafer is not perfectly round. It must be noted that less rotation was observed during tracking than under steady-state.

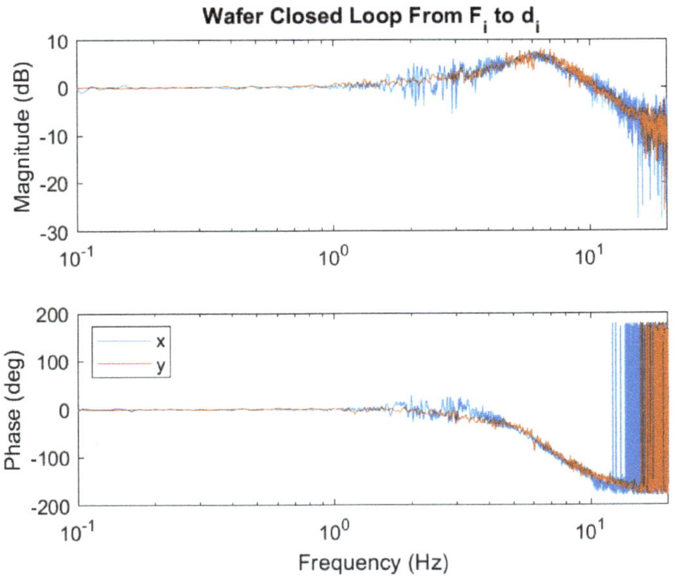

Figure 25. The closed loop frequency response of the Flowerbed—T from the reference position to the measured position for the translations.

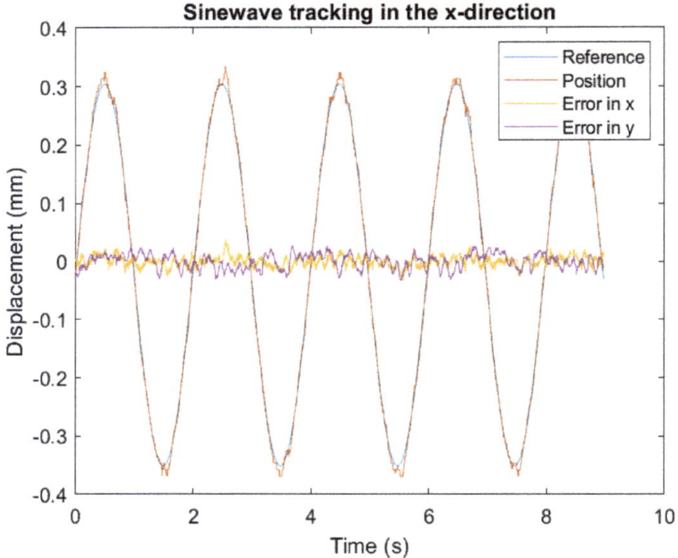

Figure 26. Tracking response of the wafer.

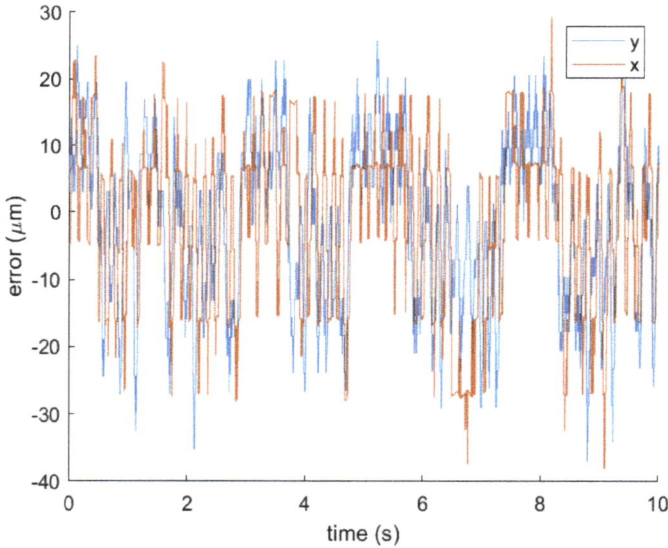

Figure 27. The steady state error during control of the wafer in x and y.

Table 5. Wafer position control-steady state performance.

DOF	RMS Error
x	12.4 μm
y	12.9 μm
Rz	3.1 mrad

7. Conclusions

The push for thinner and larger wafers has substantially increased the need for contactless handling of the substrates. The Flowerbed and other concepts were developed for this purpose using an air-bearing concept. This paper has presented an air-bearing contactless wafer handling system where the wafer's position is sensed without any alterations to the wafer.

The performance of the inner loop has been improved from 233 Hz to 300 Hz by improving the controller. This is an improvement of almost 30 percent compared to previously published results [13]. The inner loop creates a stable platform for the outer loop controller which controls the position of the wafer. External disturbances are handled in the inner loop, since the air does not transmit the vibrations from the floor.

With the use of three charge-coupled devices (CCD) the position of the wafer could be measured by detecting the edge of the wafer. This is done without any alteration of the wafer. The CCD sensors have a resolution of 5.5 μm.

Due to the unforeseen lowering of the pneumatic gain than attained in previous studies, the bandwidth for control has not been improved. However, the wafer has been controlled using the new sensor's design, resulting in a bandwidth of 10 Hz. The wafer can be positioned with a high accuracy of 13 μm RMS, which is under 3 resolution counts of the sensor. This showcases the high precision positioning capability of the Flowerbed.

Further improvement of the inner loop could be achieved without redesign of the actuator. This improvement can be made by laminating the mover such that the effect from the eddy currents is removed. If lamination is not an option switching to flux feedback could remove the effect from the eddy currents, which are lowering the phase and thus the bandwidth. Additionally, decoupling filters could be used to improve the closed-loop sensitivity function of the inner loop. However, even in the current state the inner loop is no longer limiting the performance of the outer loop.

The resolution of the CCD could be improved to sub-micrometre resolution using sub-pixel interpolation techniques. Additionally, using improved data acquisition hardware it is possible to increase the sampling frequency of the current CCD from 320 Hz to 1.2 kHz.

Author Contributions: Conceptualization, all; methodology, R.H., N.S. and S.H.H.; software, R.H. and N.S.; validation, R.H., N.S., S.H.H. and R.A.J.v.O.; investigation, R.H.; writing—original draft preparation, R.H. and N.S.; writing—review and editing, all; supervision, R.A.J.v.O. All authors have read and agreed to the published version of the manuscript.

Funding: This research received no external funding.

Conflicts of Interest: The authors declare no conflict of interest.

References

1. Fraunhofer ISE. *Photovoltaics Report*; Fraunhofer ISE: Freiburg, Germany, 2018.
2. Carey, S.; Knotter, D.M.; Ooms, E.; Boersma, J.; van der Sar, E.; Cop, R.; Gerritzen, W.; van Zadelhoff, H.; Bouma, H. Yield Impact of Backside Metal-Ion Contamination. In *Solid State Phenomena*; Trans Tech Publications: Stafa-Zurich, Switzerland, 2012; Volume 187.
3. Wesselingh, J. Contactless Positioning Using an Active Air Film. Ph.D. Thesis, Delft University of Technology, Delft, The Netherlands, 2011. Available online: https://repository.tudelft.nl/islandora/object/uuid%3A4f90d3b0-1698-4885-b4b4-bdec9d886893 (accessed on 17 August 2021)
4. Vuong, P.H. Air-Based Contactless Actuation System for Thin Substrates. Ph.D. Thesis, Delft University of Technology, Delft, The Netherlands, 2016. Available online: https://repository.tudelft.nl/islandora/object/uuid%3A2d375f1b-3857-4c03-87e8-cb0fc45f3f13 (accessed on 17 August 2021).
5. Verbruggen, N. Air-Based Contactless Positioning of Thing Substrates. Master's Thesis, Delft University of Technology, Delft, The Netherlands, 2018. Available online: https://repository.tudelft.nl/islandora/object/uuid%3A42b0d71c-3055-41a8-8234-5acc8319f378 (accessed on 17 August 2021)
6. Vagher, E.P. Contactless Passive Transport of Thin Solar Cells. Master's Thesis, Delft University of Technology, Delft, The Netherlands, 2016. Available online: https://repository.tudelft.nl/islandora/object/uuid%3A1386cbd2-eca4-44a1-9dce-b8ec2083176b (accessed on 17 August 2021).

7. Snieder, J. Development of an Air-Based Contactless Transport Demonstrator. Master's thesis, Delft University of Technology, Delft, The Netherlands, 2017. Available online: https://repository.tudelft.nl/islandora/object/uuid%3A870adc6e-2f78-468c-a6c2-079eaec49226 (accessed on 17 August 2021).
8. Voorrips, Y. The Design of a Distributed Sensing System for Contactless Substrate Transport. Master's Thesis, Delft University of Technology, Delft, The Netherlands, 2017. Available online: https://repository.tudelft.nl/islandora/object/uuid%3A97c923e7-09d3-439b-87f2-e88d786d71b5 (accessed on 17 August 2021).
9. Laurent, G.J.; Hyungpil, M. A survey of non-prehensile pneumatic manipulation surfaces: principles, models and control. *Intell. Serv. Robot.* **2015**, *8*, 151–163. [CrossRef]
10. Spaan-Burke, T.M.; van Dam, T.; Overschie, P.; Spaan, H.A. Contactless Shape Manipulation of Thin Substrates Using Air Bearing Table. In Proceedings of the 2017 ASPE Annual Meeting, Charlotte, NC, USA, 29 October–3 November 2017
11. Kim, J.-H. Design of the Air Pressure Pick-up Head for Non-Contact Wafer Gripper. *J. Korean Soc. Manuf. Technol. Eng.* **2012**, *21*, 401–407.
12. Reinhart, G.; Heinz, M.; Stock, J.; Zimmermann, J.; Schilp, M.; Zitzmann, A.; Hellwig, J. Non-contact handling and transportation for substrates and microassembly using ultrasound-air-film-technology. In Proceedings of the 2011 IEEE/SEMI Advanced Semiconductor Manufacturing Conference, Saratoga Springs, NY, USA, 16–18 May 2011
13. Jansen, R. Actuation System for the FLowerbed. Master's Thesis, Delft University of Technology, Delft, The Netherlands, 2018. Available online: https://repository.tudelft.nl/islandora/object/uuid%3Ab4a3a5da-c512-40ed-b631-cf2d6e638ce6 (accessed on 17 August 2021).
14. Krijnen, M.E. Control System Design for a Contactless Actuation System. Master's Thesis, Delft University of Technology, Delft, The Netherlands, 2016. Available online: https://repository.tudelft.nl/islandora/object/uuid%3Abef72c5e-7ca1-41eb-85ba-fc6fc59a6ce2 (accessed on 17 August 2021).
15. Krijnen, M.E.; van Ostayen, R.A.; HosseinNia, H. The application of fractional order control for an air-based contactless actuation system. *ISA Trans.* **2018**, *82*, 172–183. [CrossRef] [PubMed]
16. Katalenic, A. Control of Reluctance Actuators for High-Precision Positioning. Ph.D. Thesis, Technische Universiteit Eindhoven, Eindhoven, The Netherlands, 2013.
17. Li, C.-M.; Xu, G.-S. Sub-pixel Edge Detection Based on Polynomial Fitting for Line-Matrix CCD Image. In Proceedings of the 2009 Second International Conference on Information and Computing Science, Manchester, UK, 21–22 May 2009; pp. 262–264. [CrossRef]
18. Xu, G.-S. Linear Array CCD Image Sub-pixel Edge Detection Based on Wavelet Transform. In Proceedings of the 2009 Second International Conference on Information and Computing Science, Manchester, UK, 21–22 May 2009; pp. 204–206. [CrossRef]
19. Fischer, J.; Pribula, O. Precise subpixel position measurement with linear interpolation of CMOS sensor image data. In Proceedings of the 6th IEEE International Conference on Intelligent Data Acquisition and Advanced Computing Systems, Prague, Czech Republic, 15–17 September 2011; pp. 500–504. [CrossRef]
20. Hagara, M.; Kulla, O. Edge detection with sub-pixel accuracy based on approximation of edge with Erf function. *Radioengineering* **2011**, *20*, 516–524.
21. Jan, F.; Tomas, R. Simple methods of edge position measurement using shadow projected on CCD sensor. *Meas. Sci. Rev.* **2003**, *3*, 37–40.

Article

Rotation Accuracy Analysis of Aerostatic Spindle Considering Shaft's Roundness and Cylindricity

Guoqing Zhang [1,2], Jianming Zheng [1,*], Hechun Yu [2], Renfeng Zhao [1], Weichao Shi [1] and Jin Wang [2]

[1] School of Mechanical and Precision Instrument Engineering, Xi'an University of Technology, Xi'an 710048, China; cims@msn.cn (G.Z.); zrf20070607@xaut.edu.cn (R.Z.); shiweichao@xaut.edu.cn (W.S.)

[2] School of Mechatronics Engineering, Zhongyuan University of Technology, Zhengzhou 451191, China; 6222@zut.edu.cn (H.Y.); 6722@zut.edu.cn (J.W.)

* Correspondence: zjm@xaut.edu.cn

Abstract: The rotation accuracy of the aerostatic spindle can easily be affected by shaft shape errors due to the small gas film clearance. Thus, the main shaft shape errors with the largest scale—that is, the roundness and cylindricity errors—are studied in this paper, and a dynamic mathematical model is established with the consideration of the roundness, cylindricity errors, and spindle speed. In order to construct the shaft model, the discrete coefficient index of the shaft radius based on roundness measurement data are proposed. Then, the simulation calculations are conducted based on the measured cylindricity data and the constructed shaft model. The calculation results are compared with the spindle rotation accuracy measured using the spindle error analyzer. The results show that the shaft with a low discrete coefficient is subjected to less unbalanced force and smaller rotation errors, as obtained by the experiment.

Keywords: rotation accuracy; aerostatic spindle; roundness; cylindricity; dispersion coefficient

Citation: Zhang, G.; Zheng, J.; Yu, H.; Zhao, R.; Shi, W.; Wang, J. Rotation Accuracy Analysis of Aerostatic Spindle Considering Shaft's Roundness and Cylindricity. *Appl. Sci.* **2021**, *11*, 7912. https://doi.org/10.3390/app11177912

Academic Editors: Terenziano Raparelli, Federico Colombo, Andrea Trivella and Luigi Lentini

Received: 27 July 2021
Accepted: 25 August 2021
Published: 27 August 2021

Publisher's Note: MDPI stays neutral with regard to jurisdictional claims in published maps and institutional affiliations.

1. Introduction

In contrast with traditional spindles, an aerostatic spindle uses compressed gas to form a gas film, which has the characteristics of high precision, high speed, and low noise. The film thickness of an aerostatic spindle is usually about 10 micrometers, which meets the high requirements regarding manufacturing errors relating to the shaft [1]. Manufacturing errors can be classified as dimensional errors and shape errors, which include roundness, cylindricity, waviness, and roughness. Shape errors of the shaft will cause the uneven distribution of the gas film in the bearing. When the shaft rotates, the thickness of the gas film in the bearing will change continuously, which will cause pressure fluctuation, and then produce an unbalanced force on the shaft. The unbalanced force will affect the stability and rotation accuracy of the spindle.

Pande et al. [2] pointed out that the shape errors of journal bearings can generally reduce the load-carrying ability and increase the flow rate. Song et al. [3] calculated the half frequency whirl phenomena and the instability threshold speeds for the bearings with shape errors. They found that the precision of the shaft is much more important than that of the bearing. Sun et al. [4] studied the effects of shape errors of a shaft-bearing system on dynamic characteristics, and results show that the rotating speed at which the fluid whip occurs increased when shape errors existed. In order to reduce the radial error motion, Cappa et al. [5] analyzed the influence of several manufacturing errors, bearing parameters and feeding geometries of an aerostatic journal bearing and suggested that the radial error motion is mainly influenced by the shape errors of the shaft. Cui et al. [6] presented research on the manufacturing errors of aerostatic porous journal bearings, and found that the circumferential waviness errors caused the obvious inhomogeneity of the flow field and the transformation of the morphology of the high-pressure region. Zhang et al. [7] established a new approximate model to predict the radial error motion of hydrostatic journal bearings, and pointed out that the main working harmonic components in roundness errors had a major influence over rotation accuracy. Wang et al. [8] proposed a numerical model using linear perturbation theory for the research on the influence of journal rotation and bearing surface waviness on the dynamic performance of aerostatic journal bearings. Lee et al. [9] examined the influence of the waviness errors of a hydrostatic journal bearing by considering the rotational and bearing installation angles, and revealed that the load-carrying capacity varied according to the amplitude and phase angle of the waviness functions. Li et al. [10] investigated the effects of different forms of surface waviness on the stability of the hydrodynamic journal bearing system by using non-linear dynamic analysis and calculation method, and the results show that the existence of the surface waviness causes the oil film of the system to wave and increases the energy loss of the system. In order to study the surface roughness effects in hydrodynamic bearings, Quiñonez et al. [11] developed an analytical solution method based on the Reynolds equation using perturbation techniques under the assumption of waviness coupled with linear superposition that can be used in wide exponential land slider bearings. Lin [12] analyzed the effect of surface roughness on the dynamic stiffness and damping characteristics of hydrostatic thrust bearings; according to the results, the mean stiffness and damping behaviours are significantly affected by the roughness pattern and the height of the roughness. Kumar et al. [13] studied the effect of stochastic roughness on performance of a Rayleigh step bearing operating under thermo-elastohydrodynamic lubrication. The results show that the directional orientation of the surface roughness could affect the performance of a bearing significantly. Based on the stochastic method and the Ng–Pan turbulent model, Zhu et al. [14] calculated the turbulent lubrication performance of a journal bearing with an isotropic rough surface. Kim et al. [15] proposed an approach with both stochastic and contact characteristics to evaluate the effects of surface roughness for a slider bearing operating under partial or boundary lubrication, and found that the influence of roughness parameters on friction and load capacity increased rapidly upon the application of asperities contact. Maharshi et al. [16] proposed a stochastic analysis of hydrodynamic journal bearings including the effect of surface roughness and development of the efficient radial basis function based uncertainty quantification algorithm. By using Galerkin's technique to solve the Reynolds equation, Rajput [17] indicated that the performance of a journal bearing system is significantly deteriorated due to the geometric imperfections.

In the above references, the waviness and roughness in the shaft shape errors are mostly studied. However, the roundness and cylindricity are the largest errors among the shaft shape errors, usually between 1 and 4 micrometers. These errors have the most obvious impact on the aerostatic spindle's rotation accuracy. Although Zhang et al. [7] studied the radial error motion of the bearing in a quasi-static state considering roundness errors of the shaft, the rotation accuracy of the aerostatic spindle will change with rotation speed, which is not considered in the present study. Therefore, it is necessary to study the dynamic rotation accuracy of the aerostatic spindle under the effects of roundness and cylindricity errors.

Therefore, a dynamic mathematical model considering roundness, cylindricity errors and spindle speed is established by using the finite difference method based on the Reynolds equation in this paper. The roundness and cylindricity errors of the shaft are measured by Taylor Hobson 585 LT cylindricity meter. The measured data are resampled and filtered and then brought into the model for calculation. Through the comparison of the simulation and the experiment, it is proved that the shaft roundness and cylindricity errors will produce an unbalanced gas film force on the shaft and affect the spindle's rotation accuracy. Based on the measurement data of roundness error, the evaluation index of the dispersion coefficient is proposed. The research shows that this discrete coefficient is a more effective index to predict the spindle's rotation accuracy compared with the roundness and cylindricity.

2. Mathematical Model

2.1. Modeling of the Aerostatic Spindle

The schematic illustration of the aerostatic journal bearing with shaft shape errors is shown in Figure 1. This bearing has two column orifice restrictors at the front and rear, and each column includes eight orifices. The compressed gas flows into the clearance between the bearing and the shaft through the orifice, and then flows out from both ends of the bearing.

Due to the shape errors of shaft surface, the film thickness of different positions in the circumferential and axial directions of the bearing is different. In Figure 1b, O_0 is the bearing center, O_1 is the shaft center, and R is the bearing radius. h is the film thickness at point A on the shaft surface that can be denoted as

$$h = R - \left| \overrightarrow{O_0O_1} + \overrightarrow{O_1A} \right| \tag{1}$$

When θ is the angle between $\overrightarrow{O_0A}$ and the x-axis, $\overrightarrow{O_0O_1} = (e_x, e_y)$, and $\overrightarrow{O_1A} = (c_{x\theta}, c_{y\theta})$, the film thickness at θ is

$$h_\theta = R - \sqrt{\left(c_{x\theta} + e_x\right)^2 + \left(c_{y\theta} + e_y\right)^2} \tag{2}$$

If the shaft is rotating clockwise at a constant angular velocity ω, then at t time,

$$\begin{cases} \theta' = \theta + \omega t + \alpha - 2N\pi \\ c_{x\theta} = x_{\theta'} \\ c_{y\theta} = y_{\theta'} \end{cases} \tag{3}$$

where α is the initial angle of the shaft, N is the integer number of rotations of the shaft, and the values of $x_{\theta'}$ and $y_{\theta'}$ come from the measurement results of the cylindricity meter.

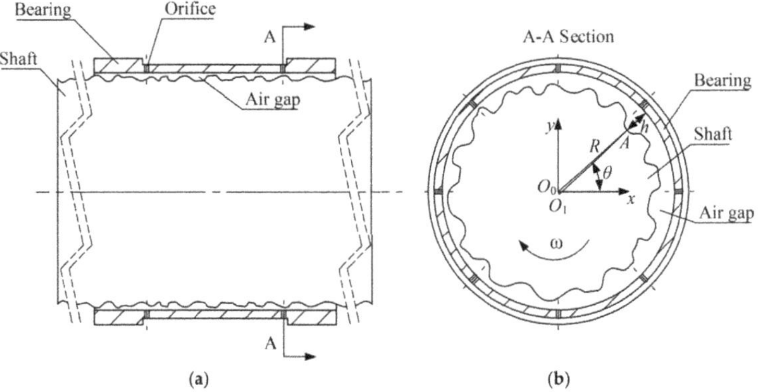

Figure 1. Illustration of the aerostatic journal bearing with shaft shape errors: (**a**) axial direction; (**b**) circumferential direction.

By combining Equations (2) and (3), the film thickness at any angle at any time can be obtained.

In order to study the pressure distribution of the gas film in the bearing, the circumferential cylindrical coordinates system is used to reconstruct the journal bearing. Figure 2 is the partial schematic illustration of aerostatic journal bearing in circumference cylindrical coordinates. The circumferential direction of bearing's inner surface is the x-axis and the y-axis points and is perpendicular to the bearing axis, and the z-axis is parallel to the bearing axis.

This spindle system incorporates the following assumptions:

1. The spindle is cooled sufficiently—that is, the bearing, shaft and gas film is isothermal;
2. There is no axial and angular movement of the shaft;
3. The gas flow is laminar.

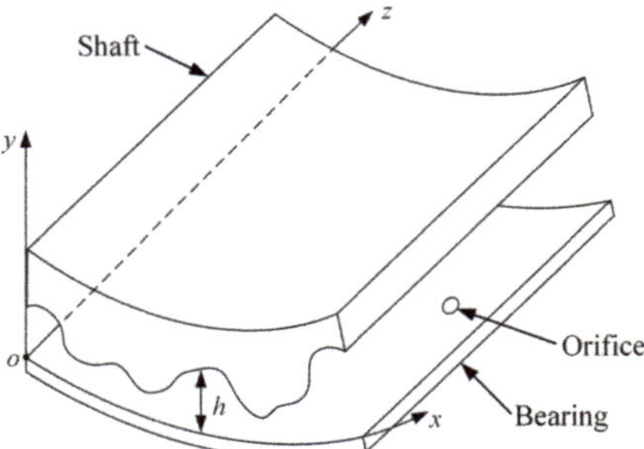

Figure 2. Circumference cylindrical coordinates.

The modified Reynolds equation for this type of aerostatic journal bearing [18] is

$$\frac{\partial}{\partial x}\left(ph^3\frac{\partial p}{\partial x}\right)+\frac{\partial}{\partial z}\left(ph^3\frac{\partial p}{\partial z}\right)=6\eta u\frac{\partial(ph)}{\partial x}+12\eta\frac{\partial(ph)}{\partial t}+12\eta p V_{in} \tag{4}$$

where p is the gas film pressure, η is the dynamic viscosity of the gas, u is the velocity of the gas in x direction, and V_{in} is the injection velocity at the orifice entrance and can be expressed by Equation (5) [19]

$$V_{in}=-\frac{P_s-p}{4\eta l}[\frac{d^2}{4}-(x-x_i)^2-(z-z_i)^2]\cdot\delta_i \tag{5}$$

where $\delta_i = 1$ at the orifice entrance, whereas it is zero elsewhere.

By employing the dimensionless parameters

$$x=x_0X, z=z_0Z, h=h_0H, t=\tau\frac{x_0}{u}, p=p_aP, \Lambda=\frac{12\eta u x_0}{h_0^2 p_a}, Q=\frac{24\eta x_0^2}{h_0^3 p_a^2}V_{in} \tag{6}$$

Equation (4) can be written as:

$$\frac{\partial}{\partial X}\left(H^3\frac{\partial P^2}{\partial X}\right)+\frac{x_0^2}{z_0^2}\frac{\partial}{\partial Z}\left(H^3\frac{\partial P^2}{\partial Z}\right)=\Lambda\frac{\partial PH}{\partial X}+2\Lambda\frac{\partial PH}{\partial \tau}+QP \tag{7}$$

2.2. Numerical Analysis

With reference to paper [20], some items in Equation (7) are rewritten as follows by using the finite difference method:

$$\frac{\partial}{\partial X}\left(H^3\frac{\partial P^2}{\partial X}\right)=\frac{H_{i+0.5,j}^3}{\Delta X^2}\left(P_{i+1,j}^2-P_{i,j}^2\right)-\frac{H_{i-0.5,j}^3}{\Delta X^2}\left(P_{i,j}^2-P_{i-1,j}^2\right) \tag{8}$$

$$\frac{\partial}{\partial Z}\left(H^3\frac{\partial P^2}{\partial Z}\right)=\frac{H_{i,j+0.5}^3}{\Delta Z^2}\left(P_{i,j+1}^2-P_{i,j}^2\right)-\frac{H_{i,j-0.5}^3}{\Delta Z^2}\left(P_{i,j}^2-P_{i,j-1}^2\right) \tag{9}$$

$$\frac{\partial PH}{\partial X}=\frac{H_{i+0.5,j}-H_{i-0.5,j}}{2\Delta X}P_{i,j}+\frac{H_{i+0.5,j}}{2\Delta X}P_{i+1,j}-\frac{H_{i-0.5,j}}{2\Delta X}P_{i-1,j} \tag{10}$$

$$\frac{\partial PH}{\partial \tau}=\frac{H_{i,j}}{\Delta\tau}P_{i,j}-\frac{H_{(\tau-\Delta\tau)(i,j)}}{\Delta\tau}P_{(\tau-\Delta\tau)(i,j)} \tag{11}$$

where $H_{(\tau-\Delta\tau)(i,j)}$ and $P_{(\tau-\Delta\tau)(i,j)}$ in Equation (11) are $H_{i,j}$ and $P_{i,j}$ in $(\tau-\Delta\tau)$ time, respectively.

By substituting Equations (8)–(11) into Equation (7), the equation is simplified as

$$E_{i,j}P_{i,j}^2 + (F_{i,j} + I_{i,j})P_{i,j} - K_{i,j} = 0 \tag{12}$$

with

$$A_{i,j} = \frac{H_{i+0.5,j}^3}{\Delta X^2} \tag{13}$$

$$B_{i,j} = \frac{H_{i-0.5,j}^3}{\Delta X^2} \tag{14}$$

$$C_{i,j} = \frac{x_0^2}{z_0^2}\frac{H_{i,j+0.5}^3}{\Delta Z^2} \tag{15}$$

$$D_{i,j} = \frac{x_0^2}{z_0^2}\frac{H_{i,j-0.5}^3}{\Delta Z^2} \tag{16}$$

$$E_{i,j} = A_{i,j} + B_{i,j} + C_{i,j} + D_{i,j} \tag{17}$$

$$F_{i,j} = \Lambda\frac{H_{i+0.5,j} - H_{i-0.5,j}}{2\Delta X} \tag{18}$$

$$G_{i,j} = \Lambda\frac{H_{i+0.5,j}}{2\Delta X} \tag{19}$$

$$H_{i,j} = -\Lambda\frac{H_{i-0.5,j}}{2\Delta X} \tag{20}$$

$$I_{i,j} = 2\Lambda\frac{H_{i,j}}{\Delta\tau} + Q \tag{21}$$

$$J_{i,j} = -2\Lambda\frac{H_{(t-\Delta\tau)(i,j)}}{\Delta\tau} \tag{22}$$

$$K_{i,j} = A_{i,j}P_{i+1,j}^2 + B_{i,j}P_{i-1,j}^2 + C_{i,j}P_{i,j+1}^2 + D_{i,j}P_{i,j-1}^2 - G_{i,j}P_{i+1,j} - H_{i,j}P_{i-1,j} - J_{i,j}P_{(t-\Delta\tau)(i,j)} \tag{23}$$

The dimensionless gas film pressure in τ time is given by

$$P_{i,j} = \frac{\sqrt{(F_{i,j} + I_{i,j})^2 + 4E_{i,j}K_{i,j}} - F_{i,j} - I_{i,j}}{2E_{i,j}} \tag{24}$$

3. Data Acquisition and Processing

In addition to the basic parameters of the spindle, it is necessary to know the initial distribution of the gas film to solve the pressure distribution using Equation (24). That is, the H of any node at $\tau = 0$ time is needed, which is very critical. It can be seen from Equations (2), (3), and (6) that in order to calculate H, the shape errors of the shaft should be measured. A cylindricity meter was used to measure the roundness and cylindricity errors for study in this paper.

3.1. Roundness and Cylindricity Errors Measurement of Shaft

In this paper, we processed 4 shafts with the same parameters and technique. The roundness and cylindricity errors of the working area of these shafts corresponding to the journal bearing was measured using the Taylor Hobson 585 LT cylindricity meter, and 21 sections of each shaft were collected. The section interval was 5 mm, and the number of sampling points was 18,000 points per section. Figure 3 shows the roundness and cylindricity error measurement process, and the cylindricity error results of shaft 1 are shown in Figure 4. The data on the left-hand side represent the runout of each section, which is equal to the roundness errors under the least square method. The data on the right side represent the measured height. The cylindricity errors of four shafts are 4.00, 4.50, 4.25, and 4.15 µm, respectively.

Figure 3. The roundness and cylindricity errors measurement process.

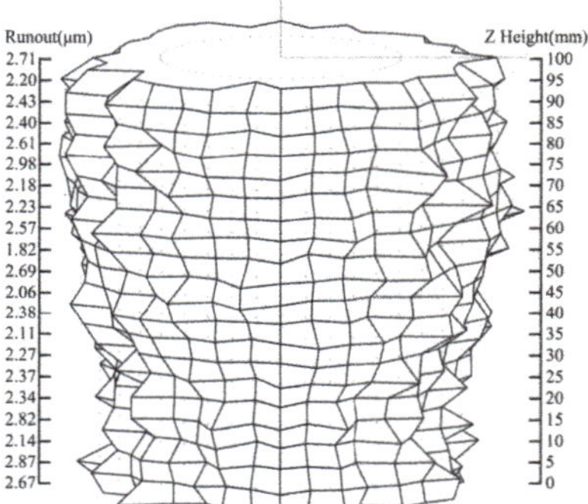

Figure 4. The cylindricity error measurement results of shaft 1.

3.2. Data Processing

Since the data collected by the cylindricity meter were based on relative coordinates, it was necessary to calibrate the measured data with the actual radius of the shaft. The calculation method is shown in Equation (25)

$$\begin{cases} x_{\theta'} = x_{c\theta'} \dfrac{R_c}{\sqrt{x_{c\theta'}^2 + y_{c\theta'}^2}} \\[2em] y_{\theta'} = y_{c\theta'} \dfrac{R_c}{\sqrt{x_{c\theta'}^2 + y_{c\theta'}^2}} \end{cases} \tag{25}$$

where $x_{c\theta'}$ and $y_{c\theta'}$ are the data collected by cylindricity meter, and R_c is the shaft radius measured by micrometer.

In the data measured by cylindricity meter, each section had 18,000 data points. The average value of each adjacent 50 data points was taken to save the computing time, and the number of data points was reduced to 360. Gaussian filter was used for roundness filtering with 150 UPR (undulations per revolution) cut-off [21]. The number of axial nodes increased from 21 to 41 by using Equation (26)

$$h_{i,k+0.5} = \frac{h_{i,k} + h_{i,k+1}}{2} \qquad (k = 1, 2, \ldots, 20) \tag{26}$$

3.3. Calculation Settings

Programs were compiled according to Equations (7)–(24); Table 1 shows the parameters of the spindle. The measured data of shaft 1, shaft 2, shaft 3, and shaft 4 were used for calculation, respectively. The initial eccentricity was $(e_x, e_y) = (0,0)$, and the speed range was [500, 10,000] r/min.

Table 1. Spindle parameters.

Parameters	Value
Bearing diameter (D/mm)	32
Bearing length (L/mm)	100
Nominal radius clearance (h_0/mm)	0.01
Orifice diameter (d/mm)	0.16
Orifice length (l/mm)	2
Column number of feeding orifices	2
Number of orifices on each column	8
Atmospheric pressure (p_{atm}/Pa)	1.013×10^5
Supplied pressure (p_s/P_{atm})	4
Gas dynamic viscosity (η/Pa·s)	18.448×10^{-6}
Shaft Material	Ti-6Al-4V
Shaft density (ρ_{shaft}/kg/m^3)	4.51×10^3

4. Results and Discussions

4.1. Simulation Results

First of all, an ideal shaft without shape errors is used for the calculation to verify the correctness of the numerical model and the simulation program; Figure 5 shows the result of dimensionless pressure distribution when the shaft rotates 180° at 6000 r/min. It can be seen that the pressure distribution on the shaft surface is very regular, which is consistent with the calculation results in reference [22]. In this ideal case, the gas film force acting on the shaft is zero.

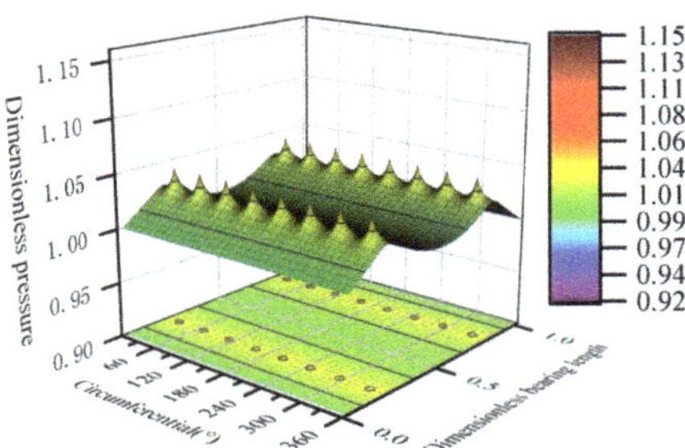

Figure 5. Dimensionless pressure distribution of ideal shaft.

Figure 6a–d show the dimensionless pressure distribution of shaft 1 at 6000 r/min when it rotates 90°, 180°, 270° and 360° with the same coordinate scales, respectively. Due to the roundness and cylindricity errors, the gas pressure on the shaft surface varies greatly. As the shaft rotates, the pressure changes obviously. Thus, the shaft will produce an unbalanced gas film force, which changes both in size and direction, and then affects the rotary accuracy of the spindle.

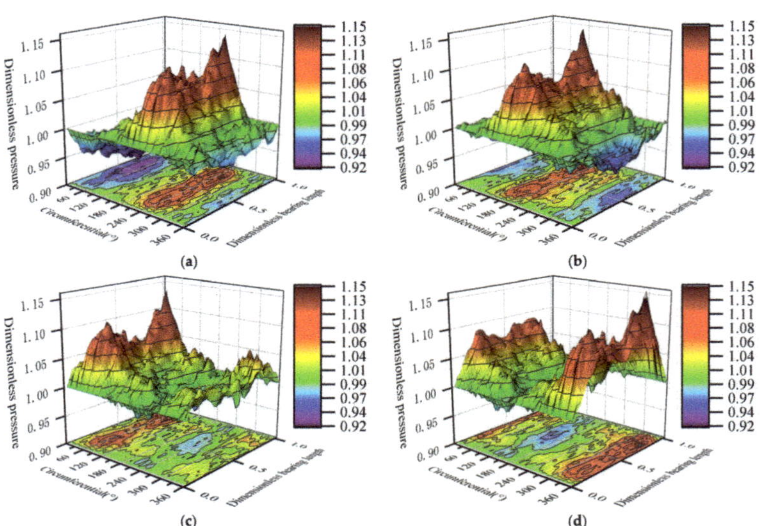

Figure 6. Dimensionless pressure distribution of shaft 1 at 6000 r/min at different angles: (**a**) 90°; (**b**) 180°; (**c**) 270°; (**d**) 360°.

Figure 7a–d are box charts of the surface pressure distribution of shaft 1, shaft 2, shaft 3, and shaft 4 at different speeds with the same coordinate scales. With the increase in rotating speed, the distribution of pressure value becomes more and more divergent. By comparing the interquartile range (IQR) and the range within 1.5IQR of the dimensionless pressure in Figure 7a–d, it can be observed that the pressure fluctuation of shaft 2 is significantly greater than that of the other three shafts.

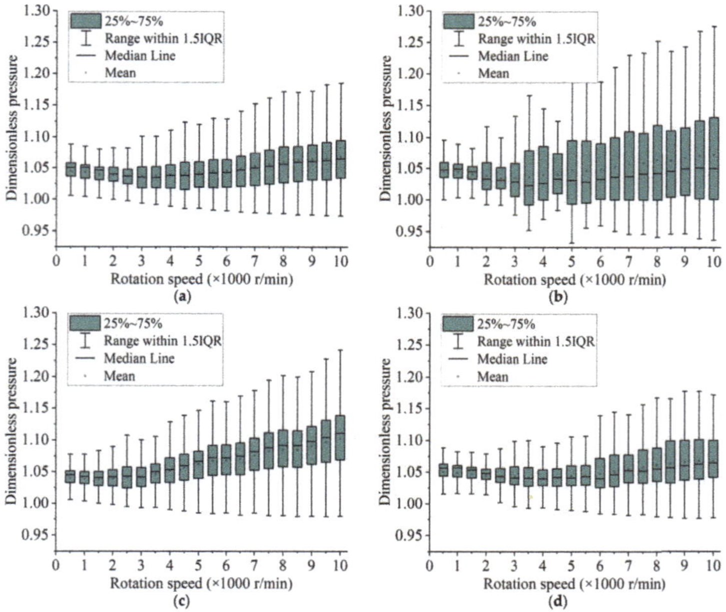

Figure 7. Box charts of surface pressure distribution: (**a**) shaft 1; (**b**) shaft 2; (**c**) shaft 3; (**d**) shaft 4.

If the pressure is known, the gas film force acting on the shaft can be calculated using Equation (27) [23]:

$$\begin{pmatrix} F_x \\ F_y \end{pmatrix} = \int_0^L \int_0^{2\pi} (p - p_{atm}) \begin{pmatrix} \cos\theta \\ \sin\theta \end{pmatrix} r d\theta dz \tag{27}$$

Figure 8a shows the gas film force (with 5 revolutions) of shaft 1 at 6000 r/min. With the rotation of the shaft, the direction and size of the force change, and the shaft adopts an unsteady state. Plotting the average force of the shaft at different speeds in Figure 8b, it can be seen that with the increase in speed, the force on the shaft increases basically. The force of shaft 2 at each speed is significantly higher than that of the other three shafts. According to the force from large to small, the order is as follows: shaft 2 > shaft 1 > shaft 4 > shaft 3.

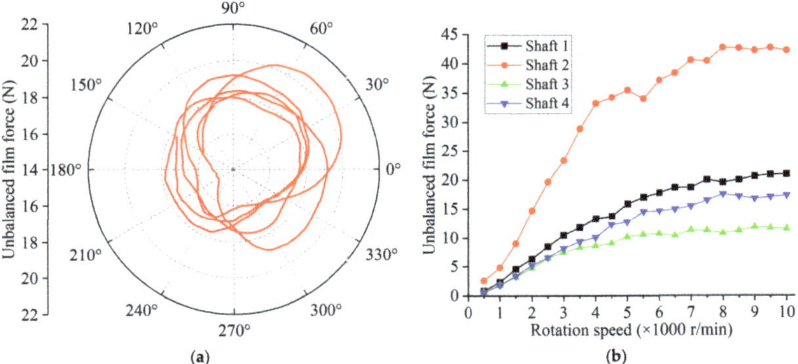

Figure 8. Gas film force of shaft: (**a**) gas film force of shaft 1 (5 revolutions) at 6000 r/min; (**b**) gas film force of shafts in different speeds.

4.2. Comparison with Experimental Results

The aerostatic spindle used in this paper to validate the simulation results is mainly used for precision milling. For ensuring the stability of continuous running, the spindle is water-cooled. The four shafts are fitted into the spindle in turn, and the bearing, motor and other parts are installed. The motor is a Kollmorgen KBMS-10X01 frameless motor with the rated speed of 18,600 r/min. The driver belongs to Sieb and Meyer's SD2S series. After two stages of drying and filtering, the gas supply pressure is set to 0.4 Mpa (same as the numerical calculation). This system is mounted on a natural granite cubic base with a side length of 200 mm. For the consistency of the measurement results, other parts and related configurations remain unchanged throughout the test, except for the replacement of the shaft.

Using sensors to measure the standard gauge installed on the spindle, the rotation errors are usually measured without load, and the track of the ideal axis of the rotation of the spindle is fitted as the basis of the analysis. Due to the small size of the spindle, there is not enough space to install the standard gauge and there is a section of finish-machined cylinder in the front of the shaft as an alternative.

The dynamic radial rotation errors were measured using the Lion Precision CPL290 spindle errors analysis instrument. The bandwidth of the sensor is 15 kHz and the resolution is 0.01 µm. The spindle speed range is set to 500 to 10,000 r/min, and the data are collected every 500 r/min. Figure 9 is the test rig of the spindle rotation error measurement, and the results of shaft 1 at 1000 r/min are shown in Figure 10. Following measurement, the data of the four shafts are plotted in Figure 11.

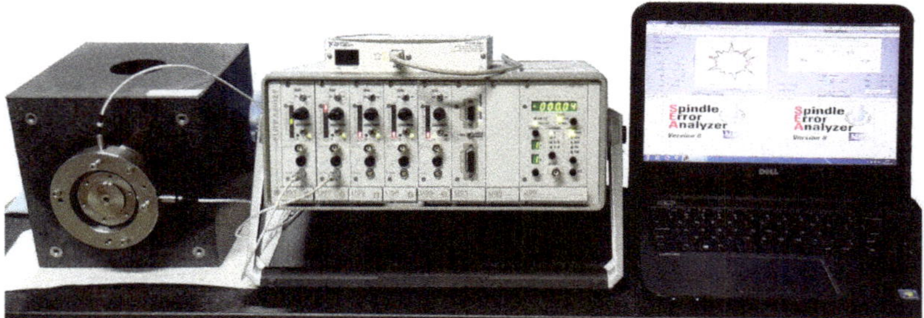

Figure 9. Test rig of spindle rotation error measurement.

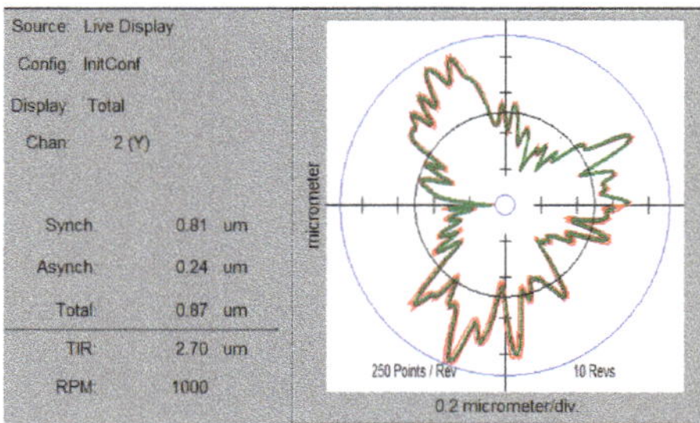

Figure 10. Result diagram of spindle radial rotation errors.

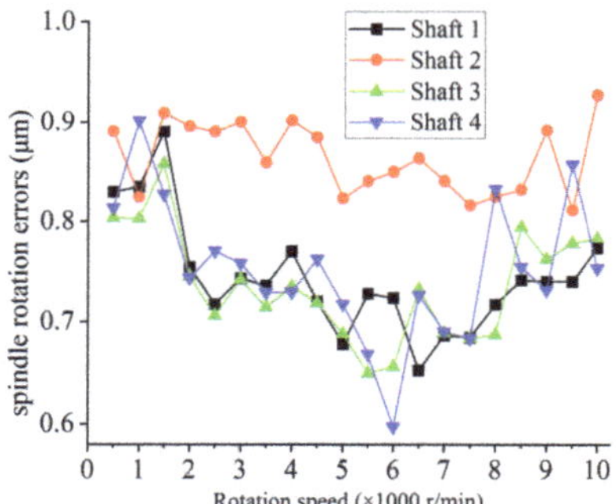

Figure 11. Spindle radial rotation errors of four shafts.

As can be seen from Figure 11, in the rotation speed range of [500, 10,000] r/min, the rotation radial errors of shaft 1, shaft 3 and shaft 4 are always close to each other, except for individual speeds. When the speed exceeds 2000 r/min, the errors of shaft 2 are obviously larger than those of the other three shafts, and the "particularity" of shaft 2 in Figure 11 is consistent with that in Figures 7 and 8b. In the analysis results of Figures 7 and 8, the original data are obtained from the measurement of the roundness and cylindricity errors of the shaft. It can be seen that there must be a correlation between the measurement results of the roundness and cylindricity errors of the shaft and the rotation errors.

The roundness errors of all shaft sections measured using the cylindricity meter are plotted, as shown in Figure 12. It can be seen that the roundness error lines of the four shafts overlap with each other and the difference is not obvious. For example, the roundness error values of section 11 of shaft 1 and section 15 of shaft 2 are both 2.69 μm, and the roundness error values of section 19 of shaft 2 and section 10 of shaft 4 are both 2.25 μm. Figure 13 is a composite drawing of roundness errors for section 19 of shaft 2 and section 10 of shaft 4. It can be seen that although the roundness errors values are the same (2.25 μm), their actual shapes are quite different. Therefore, the roundness error values cannot reflect the shape of the shaft.

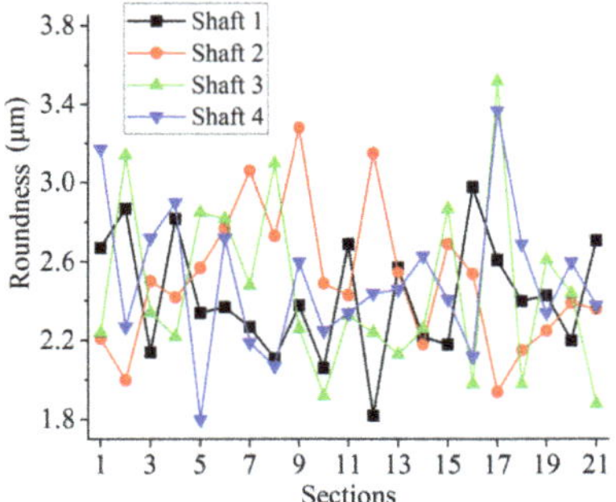

Figure 12. Roundness errors of four shafts.

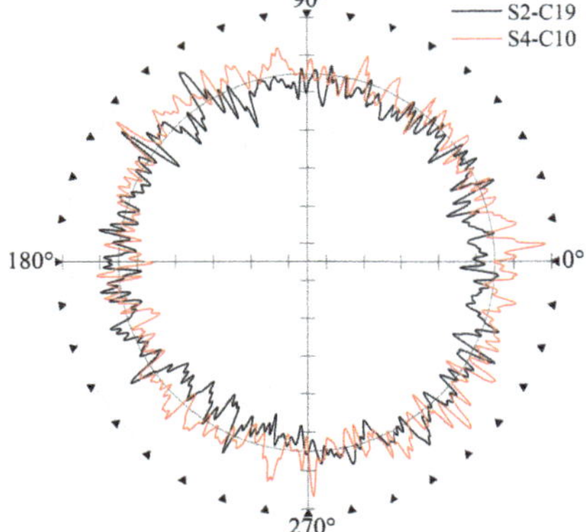

Figure 13. Composite drawing of roundness error for section 19 of shaft 2 and section 10 of shaft 4.

4.3. Evaluation with Dispersion Coefficient

The shape of the shaft determines the gas film thickness between the shaft and the matched bearing. The more uniform the film thickness, the smaller the pressure fluctuation in the shaft; therefore, the discrete coefficient index of shaft radius can be used to quantify the section shape of the shaft. The calculation is conducted by means of Equation (28):

$$V_S = \frac{1}{r_0} \sqrt{\frac{1}{n-1} \sum_{i=1}^{n} (r_i - r_0)^2} \tag{28}$$

where n is the node number of single section, r_i is the radius of the shaft at the i-th node, and r_0 is the average radius of the shaft in the current section.

The radius dispersion coefficients of all sections are calculated using Equation (28) and the results are plotted in Figure 14. It can be seen that the dispersion coefficients of all sections of shaft 2 are significantly larger than those of the other three shafts, which indicates that there are great fluctuations in radius for each section of shaft 2. The particularity results here are consistent with those shown in Figures 7, 8 and 11. Although the roundness errors of section 19 of shaft 2 and section 10 of shaft 4 are identical, the differences in the dispersion coefficients shown in Figure 14 are significant. Therefore, it is not reliable to infer the shaft's rotational accuracy based on only its roundness and cylindricity errors.

To further study this problem, sections with similar roundness errors (see Table 2) and similar discrete coefficients (see Table 3) are found from all cross sections of four shafts to construct the ideal shafts (Figure 15 is a schematic diagram), and then the unbalanced gas film forces are analyzed in Figure 16. Because the shaft is constructed with a single section, the roundness errors are equal to the cylindricity errors. Figure 16a,b have the same coordinate scales to ensure the comparability of data.

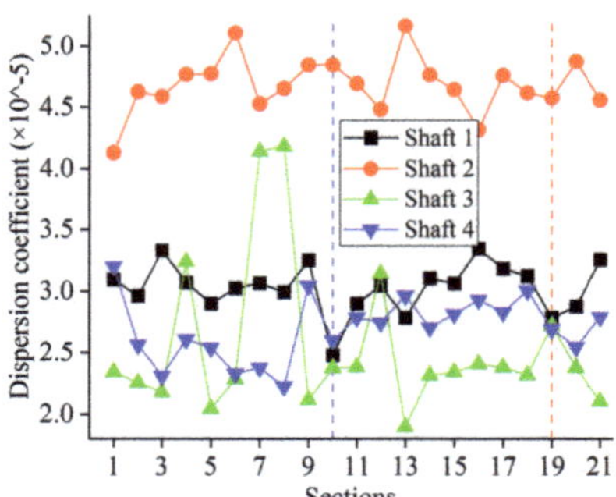

Figure 14. Dispersion coefficients of four shafts.

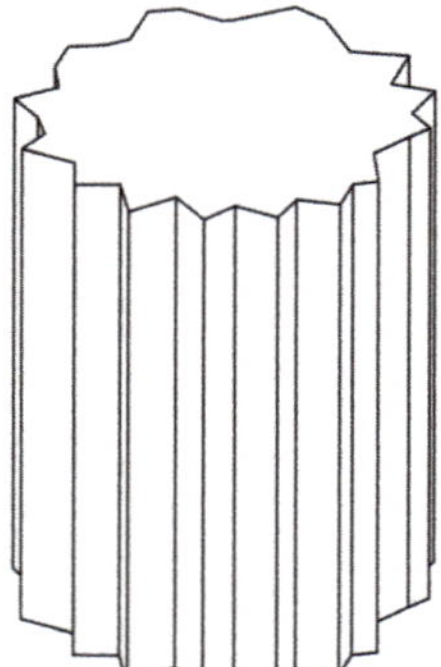

Figure 15. Schematic diagram of an ideal shaft constructed by single section.

Table 2. Sections with similar roundness.

Shaft-Cross Section	S1-C7	S2-C19	S3-C1	S4-C10
Roundness (μm)	2.27	2.25	2.24	2.25
Dispersion Coefficient (×10^{-5})	3.07	4.57	2.34	2.60

Table 3. Sections with similar discrete coefficients.

Shaft-Cross Section	S1-C2	S1-C8	S4-C13	S4-C18
Roundness (μm)	2.87	2.11	2.46	2.69
Dispersion Coefficient (×10^{-5})	2.96	2.99	2.97	3.01

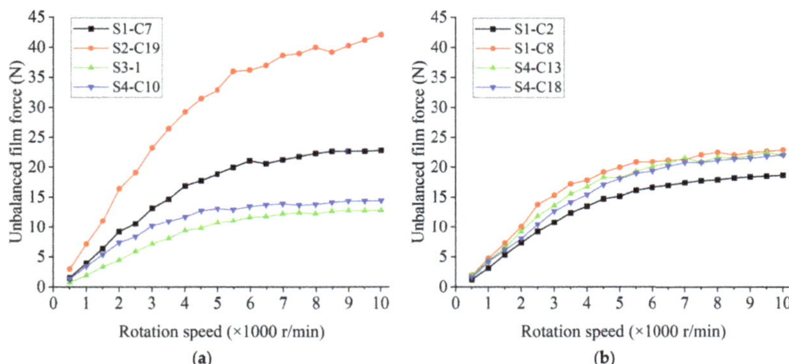

Figure 16. Unbalanced gas film force of ideal shaft: (**a**) ideal shaft with similar roundness; (**b**) ideal shaft with similar discrete coefficients.

It can be observed from Figure 16a that, although these shafts have the same roundness, the unbalanced film forces of these four shafts are quite different. The order of the force from large to small is section 19 of shaft 2 > section 7 of shaft 1 > section 10 of shaft 4 > section 1 of shaft 3, which is consistent with the order of the discrete coefficients in Table 2. As presented in Figure 16b, for the similar dispersion coefficient, the forces of the four shafts are generally similar, the lowest of which is for section 2 of shaft 1. In Table 3, the roundness errors of section 2 of shaft 1 is the largest among the four ideal shafts, but its discrete coefficient is the smallest. It can be proved that, compared with roundness errors, the discrete coefficient can be used more reliably to predict the unbalanced film force generated by the shaft due to the shape errors. In addition, when the results are combined with the data in Tables 2 and 3, it can be seen that if the roundness is similar, the discrete coefficient may be very different, but if the discrete coefficient is similar, the roundness difference is not large, so the discrete coefficient can be used to evaluate the processing quality of the shaft.

5. Conclusions

In this paper, the roundness and cylinder error data of the shaft are collected by the Taylor Hobson cylinder meter as the data source of FDM simulation. The influence of shaft shape errors on spindle rotation accuracy is verified by comparing the simulation results with experimental results. The following conclusions have been drawn.

1. Because of the errors of roundness and cylindricity in the shaft, the film thickness inside the spindle will be different at different places, resulting in an uneven distribution of film pressure. With the rotation of the shaft, the pressure of the gas film will keep changing, resulting in an unbalanced film force, which will affect the stability of the spindle;
2. The errors of roundness and cylindricity of the shaft can not adequately reflect the distribution of film thickness inside the spindle. Shafts with similar errors may have large differences in unbalanced film force and rotation errors;
3. The dispersion coefficient reflects the fluctuation of the shaft radius. Shafts with similar discrete coefficients will not demonstrate much difference in their roundness error values, and the unbalanced film forces acting on the shaft during rotation are close to each other. Compared with roundness and cylindricity errors, the discrete coefficient is a better index to predict the spindle rotation accuracy. Therefore, during the design and manufacturing process of the spindle, the shaft radius dispersion coefficient should be controlled and measured for better spindle rotation accuracy.

Author Contributions: Conceptualization, G.Z.; methodology, J.Z.; software, G.Z. and H.Y.; validation, R.Z. and W.S.; formal analysis, J.Z. and H.Y.; investigation, J.Z. and H.Y.; data curation, G.Z. and J.W.; writing—original draft preparation, G.Z.; writing—review and editing, J.Z. and J.W.; visualization, R.Z. and W.S.; funding acquisition, J.Z. and H.Y. All authors have read and agreed to the published version of the manuscript.

Funding: This research was funded by the National Natural Science Foundation of China, grant number 51875586; training plan for young backbone teachers in colleges and universities of Henan province, grant number 2018GGJS105.

Institutional Review Board Statement: Not applicable.

Informed Consent Statement: Not applicable.

Data Availability Statement: The data presented in this study are available on reasonable request from the corresponding author.

Conflicts of Interest: The authors declare no conflict of interest.

References

1. Gao, Q.; Chen, W.; Lu, L.; Huo, D.; Cheng, K. Aerostatic bearings design and analysis with the application to precision engineering: State-of-the-art and future perspectives. *Tribol. Int.* **2019**, *135*, 1–17. [CrossRef]
2. Pande, S.; Somasundaram, S. Effect of manufacturing errors on the performance of aerostatic journal bearings. *Wear* **1981**, *66*, 145–156. [CrossRef]
3. Song, M.; Azam, S.; Jang, J.; Park, S.S. Effect of shape errors on the stability of externally pressurized air journal bearings using semi-implicit scheme. *Tribol. Int.* **2017**, *115*, 580–590. [CrossRef]
4. Sun, F.; Zhang, X.; Wang, X.; Su, Z.; Wang, D. Effects of Shaft Shape Errors on the Dynamic Characteristics of a Rotor-Bearing System. *J. Tribol.* **2019**, *141*. [CrossRef]
5. Cappa, S.; Reynaerts, D.; Al-Bender, F. Reducing the radial error motion of an aerostatic journal bearing to a nanometre level: Theoretical modelling. *Tribol. Lett.* **2014**, *53*, 27–41. [CrossRef]
6. Cui, H.; Wang, Y.; Yue, X.; Huang, M.; Wang, W.; Jiang, Z. Numerical analysis and experimental investigation into the effects of manufacturing errors on the running accuracy of the aerostatic porous spindle. *Tribol. Int.* **2018**, *118*, 20–36. [CrossRef]
7. Zhang, P.; Chen, Y.; Liu, X. Relationship between roundness errors of shaft and radial error motions of hydrostatic journal bearings under quasi-static condition. *Precis. Eng.* **2018**, *51*, 564–576. [CrossRef]
8. Wang, X.; Xu, Q.; Huang, M.; Zhang, L.; Peng, Z. Effects of journal rotation and surface waviness on the dynamic performance of aerostatic journal bearings. *Tribol. Int.* **2017**, *112*, 1–9. [CrossRef]
9. Lee, S.M.; Lee, D.W.; Ha, Y.H.; Lee, S.J.; Hwang, J.H.; Choi, Y.H. A study on the influence of waviness error to a hydrostatic bearing for a crankshaft pin turner. *Tribol. Trans.* **2013**, *56*, 1077–1086. [CrossRef]
10. Li, B.; Zhou, D.; Xu, W.; Zhang, Y. Effect of Surface Waviness on Stability of Hydrodynamic Journal Bearing Systems. *J. Mech. Eng.* **2019**, *55*, 51–59. [CrossRef]
11. Quiñonez, A.F.; Morales-Espejel, G. Surface roughness effects in hydrodynamic bearings. *Tribol. Int.* **2016**, *98*, 212–219. [CrossRef]
12. Lin, J.R. Surface roughness effect on the dynamic stiffness and damping characteristics of compensated hydrostatic thrust bearings. *Int. J. Mach. Tools Manuf.* **2000**, *40*, 1671–1689. [CrossRef]
13. Kumar, R.; Azam, M.S.; Ghosh, S.K. Influence of stochastic roughness on performance of a Rayleigh step bearing operating under Thermo-elastohydrodynamic lubrication considering shear flow factor. *Tribol. Int.* **2019**, *134*, 264–280. [CrossRef]
14. Zhu, S.; Sun, J.; Li, B.; Zhao, X.; Wang, H.; Teng, Q.; Ren, Y.; Zhu, G. Stochastic models for turbulent lubrication of bearing with rough surfaces. *Tribol. Int.* **2019**, *136*, 224–233. [CrossRef]
15. Kim, M.; Lee, S.M.; Lee, D.W.; Park, S.; Kim, S. Tribological effects of a rough surface bearing using an average flow analysis with a contact model of asperities. *Int. J. Precis. Eng. Manuf.* **2017**, *18*, 99–107. [CrossRef]
16. Maharshi, K.; Mukhopadhyay, T.; Roy, B.; Roy, L.; Dey, S. Stochastic dynamic behaviour of hydrodynamic journal bearings including the effect of surface roughness. *Int. J. Mech. Sci.* **2018**, *142–143*, 370–383. [CrossRef]
17. Rajput, A.K.; Sharma, S.C. Combined influence of geometric imperfections and misalignment of journal on the performance of four pocket hybrid journal bearing. *Tribol. Int.* **2016**, *97*, 59–70. [CrossRef]
18. Pierart, F.G.; Santos, I.F. Active lubrication applied to radial gas journal bearings. Part 2: Modelling improvement and experimental validation. *Tribol. Int.* **2016**, *96*, 237–246. [CrossRef]
19. Morosi, S.; Santos, I.F. On the modelling of hybrid aerostatic-gas journal bearings. *Proc. Inst. Mech. Eng. Part J J. Eng. Tribol.* **2011**, *225*, 641–653. [CrossRef]
20. Wang, X.; Xu, Q.; Wang, B.; Zhang, L.; Yang, H.; Peng, Z. Effect of surface waviness on the static performance of aerostatic journal bearings. *Tribol. Int.* **2016**, *103*, 394–405. [CrossRef]
21. Muralikrishnan, B.; Raja, J. *Computational Surface and Roundness Metrology*; Springer Science & Business Media: Berlin/Heidelberg, Germany, 2008.

22. Lo, C.Y.; Wang, C.C.; Lee, Y.H. Performance analysis of high-speed spindle aerostatic bearings. *Tribol. Int.* **2005**, *38*, 5–14. [CrossRef]
23. Morosi, S.; Santos, I.F. Active lubrication applied to radial gas journal bearings. Part 1: Modeling. *Tribol. Int.* **2011**, *44*, 1949–1958. [CrossRef]

applied sciences

Article

Porous Gas Journal Bearings: An Exact Solution Revisited and Force Coefficients for Stable Rotordynamic Performance

Luis San Andrés [1], Jing Yang [1,*] and Andrew Devitt [2]

[1] J. Mike Walker '66 Department of Mechanical Engineering, Texas A&M University, College Station, TX 77843, USA; lsanandres@tamu.edu

[2] New Way Air Bearings, Aston, PA 19014, USA; DDevitt@newwayairbearings.com

* Correspondence: yangjing@tamu.edu

Abstract: Having come of age, gas film bearings enable high-speed oil-free (micro) rotating machinery with gains in efficiency and reliability, longer maintenance intervals, and a reduction in contaminants released to the atmosphere. Among gas bearing types, porous surface gas bearings (PGBs) have proven successful for 50+ years and presently are off-the-shelf mechanical elements. This paper reviews the literature on PGBs since the 1970s and reproduces an exact solution for the performance of cylindrical PGBs. Both the analytical model and an accompanying finite-element (FE) model predict the performance for two PGBs, a commercially available 76 mm diameter bearing and a smaller 25 mm diameter laboratory unit whose experimental performance is available. As expected, the FE model results reproduce the analytical predictions obtained in a minuscule computing time. For a set external supply pressure, as the radial clearance increases, the flow rate through the bearing grows until reaching a peak magnitude. The PGB load capacity is a fraction of the product of the set pressure difference ($p_S - p_a$) and the bearing projected area with a significantly large centering static stiffness evolving over a narrow region of clearances. Operation with shaft speed enhances the bearing load capacity; however, at sufficiently high speeds, significant magnitude cross-coupled forces limit the stable operation of a PGB. At constant operating shaft speed, as the whirl frequency grows, the bearing effective stiffness (K_{eff}) increases, while the effective damping (C_{eff}) becomes positive for whirl frequencies greater than 50% shaft speed. Similar to a plain hydrodynamic journal bearing, the PGB is prone to a half-frequency whirl, albeit the system natural frequency can be high, mainly depending on the external supply pressure. In essence, for the cases considered, PGBs are linear mechanical elements whose load capacity is proportional to the journal eccentricity.

Keywords: gas bearings; porous materials; rotordynamics; force coefficients

Citation: Andrés, L.S.; Yang, J.; Devitt, A. Porous Gas Journal Bearings: An Exact Solution Revisited and Force Coefficients for Stable Rotordynamic Performance. *Appl. Sci.* **2021**, *11*, 7949. https://doi.org/ 10.3390/app11177949

Academic Editors: Terenziano Raparelli, Andrea Trivella, Luigi Lentini and Federico Colombo

Received: 5 July 2021
Accepted: 26 August 2021
Published: 28 August 2021

Publisher's Note: MDPI stays neutral with regard to jurisdictional claims in published maps and institutional affiliations.

1. Introduction

Gas bearings support oil-free high-speed turbomachinery with significant savings in drag power losses and a reduction in system footprint [1]. Bearings constructed with a layer of porous material offer an alternative to other bearing types, such as orifice compensated bearings and foil bearings [2]. Porous surface gas bearings (PGBs) are commercially available for industrial use, in particular in linear guide systems and in the health imaging industry [3].

The contemporary approach to the modeling of PGBs relies on the numerical solution of the flow equations, using to advantage desktop computational resources [4]. However, the expedience in obtaining numerical predictions often dispenses the flow physics, avoids dimensional analysis, and disregards the effects of physical parameters that carry information on limiting solutions. Hence, this paper carefully reviews the abundant published literature on PGBs and recreates elegant theoretical analyses developed in the 1970s. The predictions from the analytical model are compared against those of a computational finite element (FE) model as well as some published experimental results.

2. The Governing Equations for a PGB and Derived Operating Parameters

The performance of PGBs is determined by their geometry (radial clearance c, length L, and diameter $D = 2R$), the gas physical properties (density ρ and viscosity μ), the magnitudes of supply pressure (p_S) and ambient pressure (p_a), the porous material permeability coefficient (κ) and the liner radial thickness (t_p), and the operating conditions of shaft angular speed (Ω and precessional frequency (ω).

Figure 1 displays a schematic diagram of a cylindrical PGB with the journal spinning with speed Ω. The film thickness is $h = (c + e_X \cos\theta + e_Y \sin\theta)$, where (e_X, e_Y) are the components of the journal static eccentricity. Consider an ideal gas with density $\rho = p/(R_g T)$, whose temperature T is constant. Then, the film pressure (p) is governed by (Gargiulo, 1979) [5]. A detailed investigation on the influence of non-zero slip flow condition and tangential flow within the porous material is out of scope in the current analysis.

$$\vec{\nabla} \cdot \left(\frac{p\, h^3}{12\mu} \vec{\nabla} p \right) = \frac{\Omega R}{2} \frac{\partial(ph)}{\partial x} + \frac{\partial(ph)}{\partial t} + \frac{\kappa}{2\mu\, t_p}(p^2 - p_S^2) \tag{1}$$

where $\vec{\nabla}$ is the vector gradient operator, μ is the gas viscosity, and p_S is the pressure of the pressurized gas supplied through the porous layer. The field p is periodic around the bearing circumference, $p_{(\theta,\,z,t)} = p_{(\theta+2\pi,z,t)}$, and on the axial ends of the bearing $p_{(\theta,\,\frac{1}{2}L,t)} = p_{(\theta,-\frac{1}{2}L,t)} = p_a$.

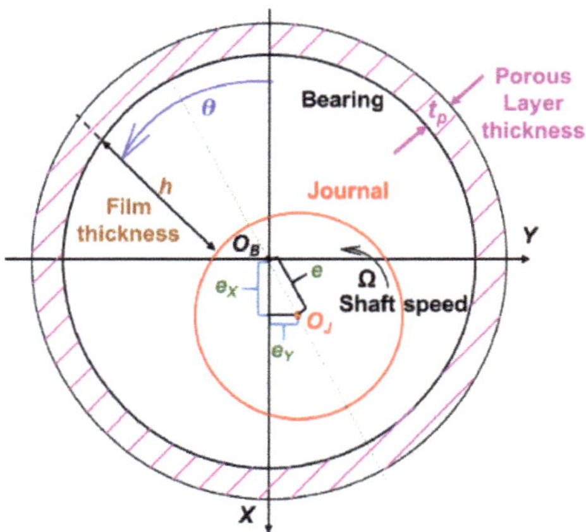

Figure 1. Schematic diagram of a cylindrical journal and bearing with porous material layer (not to scale). Coordinate system noted for reference.

Note the supplied mass flow rate through the porous material follows Darcy's law [6–8].

$$\dot{m} = \frac{\rho}{\mu} \frac{\kappa}{t_p}(p_S - p) \sim \frac{1}{2} \frac{1}{\mu\, R_g T} \frac{\kappa}{t_p} \left(p_S^2 - p^2 \right) \tag{2}$$

Flow unsteadiness, circumferential flow through the porous media, and inertial effects due to the volume in the porous material are not particularly important in modern commercial products such as carbon graphite [3,9,10].

The PGB geometry, porous material properties, and operating conditions combine to produce three fundamental (dimensionless) parameters,

$$\Lambda_\kappa = \frac{12\kappa}{t_p\, c} \left(\frac{R}{c} \right)^2 ;\; \Lambda_\Omega = 6\frac{\mu\, \Omega}{p_S} \left(\frac{R}{c} \right)^2 ;\; \Lambda_\omega = 12\frac{\mu\, \omega}{p_S} \left(\frac{R}{c} \right)^2 , \tag{3}$$

known as the flow feeding number, the speed number, and the frequency number, respectively. Clearly $p_S > p_a$ for the pressurized gas to flow through both the porous layer and then through the film clearance to exit at the bearing sides. Most PGBs have a large slenderness ratio, $L/D > 1$, hence giving enough surface area for the pressure field to produce a significant bearing load.

The Λ_Ω aerodynamic effects due to shaft rotation dominate over aerostatic effects ($\sim p_S$). Similarly, squeeze film effects (high frequency) dominate for $\Lambda_\omega > 1$. Incidentally, the porosity parameter Λ_κ relates the flow conductance across the porous layer (κ/t_p) to the conductance through the film, which is proportional to (c^3/R^2).

3. An Appraisal of the Past Literature

The archival literature on PGBs is abundant with a sound beginning in the early 1960s and continuous developments through the 1980s. Only in the 2000s did commercial applications abound, in particular in the flat panel manufacturing industry, semiconductor manufacturing industry, coordinate measuring machines, computed tomography imaging machines, and precision machine tools and spindles [3].

From 1965 to 1967, Sneck et al. publish three seminal papers [6–8] to introduce the theoretical framework for the analysis of gas lubricated PGBs. The authors derived analytical solutions for the flow rate and load capacity of PGBs under aerostatic conditions and produced experimental verification for both the load capacity as well as the flow coefficients. The authors noted the importance of surface roughness that modifies the PGB nominal clearance to produce agreement with the measured load and flow. The later reference [8] noted that shaft surface speed aids to increase the load capacity of PGBs and presents ultimate solutions for operation under very large speed numbers ($\Lambda_\Omega \gg 1$). Sneck et al. [6–8] did not study the effect of frequency on the bearing stiffness coefficients, and altogether ignored damping coefficients.

At about the same time, in 1968, Mori et al. [11,12] developed a similar solution, albeit for an incompressible fluid, and produced predictive formulas for load capacity and gas flow that agreed well with experimental results for finite length PGBs ($L/D \sim 1/2$). Both Sneck et al. and Mori et al. found that the PGB load capacity was a fraction of $W^* = ((p_S - p_a) \times (L\,D))$, i.e., the product of the pressure difference ($p_S - p_a$) times the bearing projected area. Having introduced an equivalent clearance for the layer of porous material, $c_\kappa = (12\,\kappa\,t_p)^{1/3}$, Mori et al. reported a maximum aerostatic load capacity $\sim(0.7 \times W^*)$ for the range (c_κ/c) ~ 0.6–1.0; hence, the practical flow feeding number $\Lambda_\kappa \sim (R/t_p)^2 \gg 1$.

A decade later (1978), Rao and Majumdar [13] presented a perturbation analysis and numerical solution to calculate the stiffness (K) and damping (C) force coefficients of aerostatic PGBs ($\Lambda_\Omega = 0$). The unique predictions showed $C < 0$ for small frequency numbers (Λ_ω). Interestingly enough, the analysis did not include the volume of the porous material, which, if large enough and under dynamic conditions, traps the gas to produce a pneumatic-hammer effect, i.e., a self-excited instability due to the absence (even negative) of effective damping [14].

Just a year later, in 1979, Gargiulo [5] presented a comprehensive analysis that included both steady-state performance parameters and the dynamic force coefficients via a perturbation analysis that produces analytical expressions as a function of Λ_Ω, Λ_ω, and Λ_κ, and includes the effect of the material porosity volume. For very small porous volumes, Cargiulo [1] reported a hardening bearing stiffness as frequency grows while the damping coefficient decreases steadily. The findings are similar to those for orifice-compensated aerostatic gas bearings, as published by Lund [15] in 1968. Large porous volumes could lead to a negative direct stiffness but not negative damping, in opposition to Rao and Majumdar's findings [13]. Under hybrid operations, i.e., operation with shaft speed ($\Lambda_\Omega > 0$), the analysis reports cross coupled stiffness and damping coefficients, growing in proportion to the shaft speed until $\Lambda_\Omega \to 10$, to then drop for operation at higher shaft speeds. Cargiulo [5] left for the future (the 21st century) the dynamic stability of PGBs operating in a hybrid mode.

In a companion paper, Cargiulo [16] described the outcome of experiments that for the most part validated the theoretical predictions, in particular the PGB load capacity and the direct force coefficients. The test bearing has $L = D = 50.4$ mm and a porous liner with thickness $t_p = 6.4$ mm and permeability $\kappa = 2.3 \times 10^{-9}$ mm^2 ($c_\kappa = 5.6$ μm). The experiments also revealed the onset of unstable behavior characterized by vibrations of the test system at its natural frequency. The tests conducted with various radial clearances, $c = 3$ to 17 μm, demonstrated that the test aerostatic PGB produced a journal displacement directly proportional to the applied load, even up to the point of journal contact with the porous liner surface. Hybrid operation, i.e., with shaft rotation ($\Lambda_\Omega > 0$), increases the PGB load capacity and its stiffnesses, direct and cross-coupled. Incidentally, the higher the supply pressure, the larger the bearing centering stiffness (and natural frequency), and hence the more likelihood of a PGB to be tailored for dynamically stable operation.

After a long hiatus, in 2006, Miyatake et al. [17] assessed the stability of a rotor supported on PGBs coated with a surface layer that was ~300 times more restrictive than that of the porous media itself. The manufacturing of surface loaded porous materials is a major breakthrough in PGBs, as they practically eliminate the potential for pneumatic hammer; see Otsu et al. [18].

In practice, PGBs are mounted in series with elastomeric supports (o-rings) to add more effective damping, albeit the soft mounts also reduce the system natural frequency. See for example the synopsis by Hwang [14]. Most recently, various publications detail the application of tilting pad PGBs to enable oil-free turbomachinery operating at high speeds; see San Andrés et al. [2,4,19,20] and Feng et al. [21–23]. Tilting pad journal bearings effectively eliminate the cross-coupled stiffness coefficient that could easily excite a hydrodynamic instability and thus extend the maximum operating shaft speed of rotor-bearing systems.

At Texas A&M University, since 2015, San Andrés et al. have been conducting measurements of the rotordynamic response of solid rotors supported on carbon-graphite PGBs: one rotor of small diameter (29 mm) and operating at a high shaft angular speed (55 krpm) [19], and another with large diameter (100 mm) and turning at 18 krpm maximum [4,20]. The rotors' surface speeds ($U_s = \Omega R$) equal 82 m/s and 94 m/s, respectively. Rotordynamic tests show stable responses, free of sub synchronous whirl frequencies. Imbalance mass induced synchronous whirl speed rotor motion amplitudes show peak magnitudes while crossing critical speeds largely determined by the pressure (p_S) supplied to the bearings. Derived from recorded amplification factors, the system damping ratio is larger than 10%, uncharacteristically high for gas bearing supported rotors. The computational analysis in [4] solves the Reynolds equation and Darcy's diffusion equation that couple the flow in the PGB film land to that through the porous layer. The model produces force coefficients that agree well with experimentally-derived effective damping and stiffness coefficients and shows that the magnitude of supply pressure (p_S), the permeability coefficient (κ), the pad-pivot stiffness, and the assembly clearance affect significantly the performance, static and dynamic, of the test bearings. In addition, just published in 2021, [2] describes an experimental campaign and the results that quantify the force coefficients of a large size, four tilting pad PGB. From dynamic load experiments conducted with shaft speeds at 6 and 9 krpm (32 and 64 m/s surface speed), the identified PGB's direct stiffness and damping coefficients are practically invariant with excitation frequency (to max 200 Hz), The pads' pivot compliance largely determines the bearing direct stiffness coefficients. The experiments also reveal that the load capacity of the bearing is mainly aerostatic, mostly determined by the supply pressure and the bearing preload assembly.

Feng at Hunan University leads an extensive research program whose objective is to develop oil-free bearing technologies applicable to high speed rotating machinery. Feng's group has effectively tackled bump-type foil GBs, metal mesh foil GBs, and most recently, PGBs since 2018. Feng et al. develop computational analyses for tilting pad PGBs [21,22] and cylindrical PGBs [23] and present model validations against their own experimental results as well as those in [20].

Other minor recent computational developments by Bohle [24] and Lawrence and Kemple [25] provide numerical steps toward modeling aerostatic PGBs while ignoring the vast prior literature reviewed above. Li et al.'s recent work on cylindrical PGBs [23] is particularly of notice, since the recorded static load performance of the test PGB agrees with numerical predictions. However, the identified test bearing force coefficients are rather unique since they follow a trend opposite to prior knowledge, both experimental and analytical [5,16]. That is, the PGB direct stiffness coefficient drops with frequency, while the direct damping coefficient increases.

The recent literature from 2018 through 2021 [2,21,24,25] focuses on building complex comprehensive computational tools with model verifications specific to a tailored configuration for oil-free machinery. The literature misses careful dimensional analysis, physical scaling, and dimensionless number representation that could help improve the selection of the numerical method and its validation.

This paper reworks the classical analyses of Gargiulo [5], Sneck [6], and Mori [11] to derive an exact solution for the ultimate flow rate and flow coefficient, static load capacity, and dynamic force coefficients of a generic cylindrical PGB. In the last stage of the analysis, the evaluation of integrals, the current development relies on modern mathematical software capable of symbolic algebra processing. The authors wish the engineering student and neophyte researcher appreciate the art of fundamental physical analysis, while staying for a short time away from the mundane routine of number crunching.

4. A Close Form Solution to the Flow and Dynamic Force Coefficients in a PGB

In reference to Figure 1, consider journal center motions with small amplitude (δe_X, δe_Y) and frequency (ω) about an equilibrium position (e_{X0}, e_{Y0}). The film thickness (h) and pressure (p) fields equal the sum of zeroth-order fields (h_0, p_0) and first-order or perturbed fields (h_σ, p_σ) $|_{\sigma = X, Y}$,

$$h = h_0 + \delta e_X \, h_X \, e^{i\omega t} + \delta e_Y \, h_Y \, e^{i\omega t} \rightarrow p = p_0 + \delta e_\sigma \, p_\sigma \, e^{i\omega t}, \sigma = X, Y \tag{4}$$

where $h_X = \cos(\theta)$ and $h_Y = \sin(\theta)$. Above, $\sigma = X, Y, i = \sqrt{-1}$.

Substituting Equation (4) into the Reynolds Equation (1) produces the zero-order equation for p_0

$$\vec{\nabla} \cdot \left(\frac{p_0 h_0^3}{12 \mu} \vec{\nabla} p_0 \right) = \frac{\Omega R}{2} \frac{\partial (p_0 h_0)}{\partial x} + \frac{\kappa}{2 \mu \, t_p} \left(p_0^2 - p_S^2 \right) \tag{5}$$

and the first-order equation

$$\vec{\nabla} \cdot \left(\frac{h_0^3}{12 \mu} \left[p_\sigma \vec{\nabla} p_0 + p_0 \vec{\nabla} p_\sigma \right] \right) = \frac{\Omega R}{2} \frac{\partial (p_0 h_\sigma + p_\sigma h_0)}{\partial x} + i \, \omega \, (p_0 h_\sigma + p_\sigma h_0) \\ - \vec{\nabla} \cdot \left[\frac{p_0 h_0^2}{4 \mu} h_\sigma \vec{\nabla} \cdot p_0 \right] + \frac{\kappa}{\mu \, t_p} p_0 \, p_\sigma \qquad (\sigma = X, Y) \tag{6}$$

Following San Andrés [26], the equations above are integrated in a typical finite element (FE) and assembled over the flow domain to produce sets of algebraic (nonlinear) equations for solution of the pressure fields; and subsequent calculation of the PGB reaction force, drag torque and power, and dynamic force coefficients. Details of the FE model are omitted for brevity.

For operation at a centered condition, $h_0 = c$, Equation (5) reduces to

$$\vec{\nabla} \cdot \left(\vec{\nabla} p_0^2 \right) = \frac{12 \, \kappa}{c^3 \, t_p} \left(p_0^2 - p_S^2 \right) \tag{7}$$

That dispenses with the effects of gas viscosity (μ) and shaft speed (Ω). Let $\psi_0 = (p_0^2 - p_S^2)$, then Equation (7) reduces to the simple ordinary differential equation,

$$\frac{d^2}{d\bar{z}^2}(\psi_0) - \Lambda_\kappa\,\psi_0 = 0. \tag{8}$$

Above, $\bar{z} = \left(\frac{z}{R}\right)$ with $\gamma^2 = \Lambda_\kappa = \frac{3\,\kappa}{t_p\,c}\left(\frac{D}{c}\right)^2$ as the feed flow parameter. The solution of Equation (8) with ambient pressure (p_a) at the bearing sides $\bar{z} = \pm\left(\frac{L}{D}\right)$ gives

$$\psi_{0(\bar{z})} = \psi_0|_{p\,=\,p_a}\frac{\cosh(\gamma\bar{z})}{\cosh\left(\gamma\frac{L}{D}\right)} \;\rightarrow\; p_{0(\bar{z})} = \sqrt{p_S^2 + (p_a^2 - p_S^2)\frac{\cosh(\gamma\bar{z})}{\cosh\left(\gamma\frac{L}{D}\right)}}. \tag{9}$$

The mass flow rate ($\dot{M}$) supplied to the bearing through the porous layer equals

$$\dot{M} = \oiint \dot{m}\,R\,d\theta\,dz = \gamma\left(\frac{\pi\,c^3}{6\,\mu}\right)\tanh\left(\gamma\frac{L}{D}\right)\frac{(p_S^2 - p_a^2)}{R_g T}. \tag{10}$$

Note that $\dot{M}$ is proportional to $\sqrt{\frac{\kappa\,c^3}{t_p}}$, the average gas density $\rho_* = \frac{1}{2}\frac{(p_S + p_a)}{R_g T}$, and the pressure difference ($p_S - p_a$). If the bearing clearance (c) is very large (as when there is no shaft inserted in the bearing), then $p = p_a$ over the flow domain, and the supplied flow rate is the largest and equal to

$$\dot{M}_{\max}_{\;c\,\to\infty} = \frac{1}{2}\frac{\kappa}{\mu\,t_p}\frac{(p_S^2 - p_a^2)}{R_g T}(\pi\,D\,L) \tag{11}$$

$$\frac{\dot{M}}{\dot{M}_{\max}}_{\;C_r\to\infty} = \frac{\tanh\left(\gamma\frac{L}{D}\right)}{\left(\gamma\frac{L}{D}\right)}. \tag{12}$$

The ratio of flows is only a function of the bearing geometry and the permeability coefficient (κ), since $\left(\gamma\frac{L}{D}\right) = \sqrt{\frac{3\kappa}{t_p c}\frac{L}{c}}$. Equation (11), known as the flow coefficient for PGBs [6,7], serves to estimate the permeability coefficient (κ) from measurements of the supplied flow ($\dot{M}_{\max}$) for various supply pressures, as will be shown later.

Incidentally, the shear drag torque (T_o) and power loss (P_{loss}) for the laminar flow in a centered journal bearing equal

$$T_o = \mu\,\Omega R\,(R/c)\,(\pi\,D\,L),\; P_{loss} = T_o\,\Omega. \tag{13}$$

Note that for an air film, the drag friction coefficient $f = T_o/(W^* R) \ll 1$.

As the analysis considers small amplitude motions about the centered condition ($e = 0$), then from $p = p_0 + (\delta e_X\,p_X + \delta e_Y p_Y)e^{i\,\omega\,t}$, deduce

$$p^2 = p_0^2 + 2\,p_0\,(\delta e_X p_X + \delta e_Y p_Y)\,e^{i\,\omega\,t};\; \psi = \left(p^2 - p_S^2\right) = \psi_0 + (\delta e_X\,\psi_X + \delta e_Y\,\psi_Y)\,e^{i\,\omega\,t}. \tag{14}$$

Hence, $\psi_X = 2p_0 p_X$; $\psi_Y = 2p_0 p_Y$. Presently, for small amplitude motions about $e = 0$, let

$$\psi_X = \psi_{Xc(z)}\cos\theta + \psi_{Xs(z)}\sin\theta,\; \psi_Y = \psi_{Yc(z)}\cos\theta + \psi_{Ys(z)}\sin\theta. \tag{15}$$

Then, write the first order Equation (6) for the perturbed pressure fields as

$$\frac{d^2}{d\bar{z}^2}\begin{bmatrix}\psi_{Xc}\\\psi_{Xs}\end{bmatrix} - \mathbf{A}\begin{bmatrix}\psi_{Xc}\\\psi_{Xs}\end{bmatrix} = \mathbf{C}_1\cosh(\gamma\bar{z}) + \mathbf{C}_2;$$
$$\frac{d^2}{d\bar{z}^2}\begin{bmatrix}\psi_{Yc}\\-\psi_{Ys}\end{bmatrix} - \mathbf{A}^T\begin{bmatrix}\psi_{Yc}\\-\psi_{Ys}\end{bmatrix} = \mathbf{C}_1\cosh(\gamma\bar{z}) + \mathbf{C}_2 \tag{16}$$

where $\mathbf{A} = \begin{bmatrix}(1+\gamma^2)+i\overline{\Lambda}_\omega & \overline{\Lambda}_\Omega\\ -\overline{\Lambda}_\Omega & (1+\gamma^2)+i\overline{\Lambda}_\omega\end{bmatrix}$,

$$\mathbf{C}_1 = \begin{bmatrix}i\,2\overline{\Lambda}_\omega - 3\,\gamma^2\\ -2\overline{\Lambda}_\Omega\end{bmatrix}\frac{\psi_{as}}{c}\frac{1}{\cosh\left(\gamma\frac{L}{D}\right)} \text{ and } \mathbf{C}_2 = \begin{bmatrix}i\,2\overline{\Lambda}_\omega\\ -2\overline{\Lambda}_\Omega\end{bmatrix}\frac{p_S^2}{c} \tag{17}$$

with $\psi_{as} = \left(p_a^2 - p_S^2\right), \overline{\Lambda}_\Omega = \frac{12\mu\,\Omega}{2\overline{p}_0}\left(\frac{R^2}{c^2}\right); \overline{\Lambda}_\omega = \frac{24\mu\,\omega}{2\overline{p}_0}\left(\frac{R^2}{c^2}\right) \tag{18}$

using average pressure $\overline{p}_0 = \frac{1}{L}\int_0^L p_0\,dz$. The boundary conditions for the perturbed fields are homogeneous, i.e., $\psi_{X(\bar{z}=\pm\frac{L}{D})} = \psi_{Y(\bar{z}=\pm\frac{L}{D})} = 0$. Then,

$$\psi_{Xc} = \psi_{Ys} = \psi_{Xs} = \psi_{Yc} = 0 \text{ at } \bar{z} = \pm\frac{L}{D}. \tag{19}$$

Note that Equation (16) reveals that $\psi_{Xc} = \psi_{Ys}$, $\psi_{Xs} = -\psi_{Yc}$.

Once found, integration of the pressure field (p) acting on the rotor surface gives the bearing reaction force (**F**) with components

$$-\begin{bmatrix}F_X\\F_Y\end{bmatrix} = \oint\left(P_0 - P_a + (\delta e_X\,P_X + \delta e_Y P_Y)\,e^{i\,\omega\,t}\right)\begin{bmatrix}\cos\theta\\\sin\theta\end{bmatrix} Rd\theta dz. \tag{20}$$

As the bearing is centered, the zeroth order pressure field does not produce a static force, i.e., **F** = 0. The first order pressure fields produce the matrix of complex dynamic stiffnesses **H**, where $H_{ij} = (K_{ij} + i\,\omega\,C_{ij})_{i,j\,=\,X,Y}$.

$$\begin{bmatrix}H_{XX}\\H_{YX}\end{bmatrix} = \begin{bmatrix}H_{YY}\\-H_{XY}\end{bmatrix} = \oint -p_X\begin{bmatrix}\cos\theta\\\sin\theta\end{bmatrix}R\,d\theta dz = -\oint\left(\frac{\psi_X}{2\,p_0}\right)\begin{bmatrix}\cos\theta\\\sin\theta\end{bmatrix}R\,d\theta dz. \tag{21}$$

Then substituting Equation (15) above leads to

$$H_{XX} = H_{YY} = -\pi R^2\int_0^{L/D}\frac{\psi_{Xc(\bar{z})}}{p_{0(\bar{z})}}\,d\bar{z}; \; H_{XY} = -H_{YX} = -\pi R^2\int_0^{L/D}\frac{\psi_{Yc(\bar{z})}}{p_{0(\bar{z})}}\,d\bar{z}. \tag{22}$$

The exact solution for the first-order fields is as follows: Let $\psi_X = \begin{bmatrix}\psi_{Xc} & \psi_{Xs}\end{bmatrix}^T$, then Equation (16) has the general solution

$$\psi_X = \psi_{XP1}\cosh(\gamma\bar{z}) + \psi_{XP2} + \psi_{XH} \tag{23}$$

where the coefficients of the particular solution are

$$\psi_{XP1} = -\left[\mathbf{A}-\mathbf{I}\gamma^2\right]^{-1}\mathbf{C}_1 = -\frac{1}{\Delta}\begin{bmatrix}1+i\overline{\Lambda}_\omega & -\overline{\Lambda}_\Omega\\ \overline{\Lambda}_\Omega & 1+i\overline{\Lambda}_\omega\end{bmatrix}\begin{bmatrix}i\,2\overline{\Lambda}_\omega - 3\,\gamma^2\\ -2\overline{\Lambda}_\Omega\end{bmatrix}\frac{1}{\cosh\left(\gamma\frac{L}{D}\right)}\frac{\psi_{as}}{c} \tag{24}$$

$$\psi_{XP2} = -\mathbf{A}^{-1}\mathbf{C}_2 = -\frac{1}{\Delta_\gamma}\begin{bmatrix}(1+\gamma^2)+i\overline{\Lambda}_\omega & -\overline{\Lambda}_\Omega\\ \overline{\Lambda}_\Omega & (1+\gamma^2)+i\overline{\Lambda}_\omega\end{bmatrix}\begin{bmatrix}i\,2\overline{\Lambda}_\omega\\ -2\overline{\Lambda}_\Omega\end{bmatrix}\frac{p_S^2}{c} \tag{25}$$

with $\Delta = \left[1 + i\overline{\Lambda}_\omega\right]^2 + \overline{\Lambda}_\Omega{}^2$ and $\Delta_\gamma = \left[(1 + \gamma^2) + i\overline{\Lambda}_\omega\right]^2 + \overline{\Lambda}_\Omega{}^2$ as determinants of the system of equations. $\psi_{XH} = \widehat{\psi}_H e^{s\bar{z}}$, the homogenous solution of Equation (16), satisfies

$$\left[\mathbf{I}\, s^2 - \mathbf{A}\right]\widehat{\psi}_H = 0 \tag{26}$$

whose roots are $x_{1,2} = (s_{1,2})^2 = (1 + \gamma^2) + i(\overline{\Lambda}_\omega \pm \overline{\Lambda}_\Omega)$, and with corresponding eigenvectors $\widehat{\psi}_{H_1} = \left[\begin{array}{cc} 1 & i \end{array}\right]^T$, $\widehat{\psi}_{H_2} = \left[\begin{array}{cc} 1 & -i \end{array}\right]^T$. Thus,

$$\psi_{XH} = \widehat{\psi}_H e^{s\bar{z}} = d_{1c}\,\widehat{\psi}_{H_{1c}}\cosh(\sqrt{x_1}\bar{z}) + d_{2c}\,\widehat{\psi}_{H_{2c}}\cosh(\sqrt{x_2}\bar{z}). \tag{27}$$

Satisfying the end condition $\psi_{X(\bar{z}=\frac{L}{D})} = 0$ leads to

$$\left[\begin{array}{c} d_{1c} \\ d_{2c} \end{array}\right] = -\left[\begin{array}{cc} \cosh\left(\sqrt{x_1}\frac{L}{D}\right) & \cosh\left(\sqrt{x_2}\frac{L}{D}\right) \\ i\cosh\left(\sqrt{x_1}\frac{L}{D}\right) & -i\cosh\left(\sqrt{x_2}\frac{L}{D}\right) \end{array}\right]^{-1}\left[\psi_{XP_1}\cosh\left(\gamma\frac{L}{D}\right) + \psi_{XP2}\right]. \tag{28}$$

Finally, the exact solution is

$$\psi_X = \left[\begin{array}{cc} \psi_{Xc} & \psi_{Xs} \end{array}\right]^T = \psi_{XP_1}\cosh(\gamma\bar{z}) + \psi_{XP2} + d_{1c}\left[\begin{array}{c} 1 \\ i \end{array}\right]\cosh(\sqrt{x_1}\bar{z}) + d_{2c}\left[\begin{array}{c} 1 \\ -i \end{array}\right]\cosh(\sqrt{x_2}\bar{z}). \tag{29}$$

Equation (22) implements the functions above, and using symbolic mathematical software, produces the force coefficient in a very short time (fractions of a second). In 1979, Gargiulo [5] produced a nearly identical analysis, albeit he used an influence coefficient method to obtain the solution of the ordinary differential equations (the authors were unware of the original solution in [5] until later in their work when searching for examples to validate their FE and analytical models. There is no excuse for the ignorance of past work).

5. PGB Aerostatic Operation

It is of interest to quantify the bearing operation under aerostatic conditions (without shaft speed), in particular to predict the PGB static stiffness (K_S). Let $\Omega = \omega = 0$, then $\overline{\Lambda}_\Omega = \overline{\Lambda}_\omega = 0$, and Equation (16) reduces to

$$\frac{d^2}{d\,\bar{z}^2}\left[\begin{array}{c} \psi_{Xc} \\ \psi_{Xs} \end{array}\right] - (1 + \gamma^2)\left[\begin{array}{c} \psi_{Xc} \\ \psi_{Xs} \end{array}\right] = \left[\begin{array}{c} -3\,\gamma^2 \\ 0 \end{array}\right]\frac{\psi_{as}}{c}\frac{\cosh(\gamma\bar{z})}{\cosh\left(\gamma\frac{L}{D}\right)} \tag{30}$$

whose solution is

$$\psi_{Xc_{(\bar{z})}} = \frac{3}{c}\gamma^2\left(p_a^2 - p_S^2\right)\left\{\frac{\cosh(\gamma\bar{z})}{\cosh\left(\gamma\frac{L}{D}\right)} - \frac{\cosh\left[(1 + \gamma^2)^{0.5}\bar{z}\right]}{\cosh\left[(1 + \gamma^2)^{0.5}\frac{L}{D}\right]}\right\}, \quad \psi_{Xs} = 0. \tag{31}$$

Since $\psi_X = 2p_0 p_X$; $\psi_Y = 2p_0 p_Y$, then

$$p_X = \frac{1}{2p_0}\left[\psi_{Xc(z)}\cos\theta\right], \quad p_Y = \frac{1}{2p_0}\left[\psi_{Ys(z)}\sin\theta\right]. \tag{32}$$

Substitute into Equation (21) to obtain the static stiffness $K_S = K_{XX_S} = K_{YY_S}$, and $K_{XY_S} = K_{YX_S} = 0$.

6. Experimental Estimation of Porous Material Permeability Coefficient (κ)

Figure 2 displays a photograph of a cylindrical PGB with physical dimensions listed in Table 1. The table also details the conditions for measurement of the supplied flow rate and estimation of κ. A product specification sheet is available in [27].

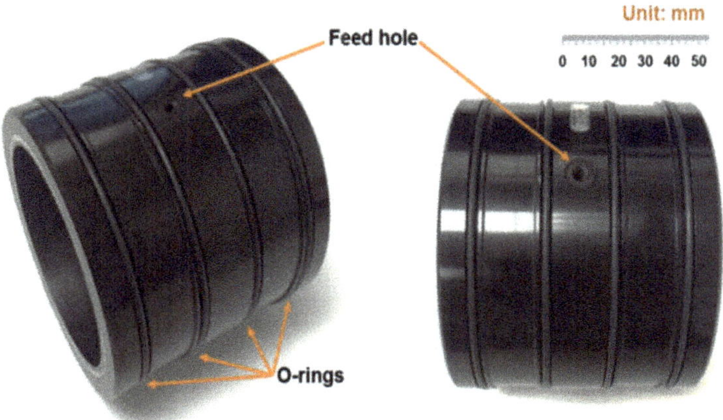

Figure 2. Photograph of a porous material cylindrical journal bearing.

Table 1. Dimensions of porous journal bearing and operating conditions.

Bearing Length, $L = 1.17\,D$	88.8 mm (3.5 inch)
inner diameter, D	76.2 mm (3.0 inch)
outer diameter, D_{out}	99.5 mm (3.9 inch)
Carbon-graphite permeability, κ	$8.2 \times 10^{-16}\ \mathrm{m}^2$
Porous layer radial thickness, t_p	2.71 mm (0.11 inch)
Equivalent clearance for porous layer, c_κ	3 μm
Supply pressure, p_S	2–8 bar
Exit pressure, p_a	1 bar
Temperature, T	294 K
Air density at (p_a, T), ρ_a	$1.2\ \mathrm{kg/m}^3$
viscosity at (p_S, T), μ	18.3×10^{-6} Pa-s
gas constant, R_g	287.05 J/(kg·K)
Parameters	For $c = 0.010$ mm
Feed flow parameter	$\Lambda_\kappa = \frac{3\,\kappa}{t_p c}\left(\frac{D}{c}\right)^2 = 5.3$
	At $p_S = 6$ bar, and $\Omega = \omega = 22{,}618$ rad/s (25 krpm)
Speed and frequency numbers	$\Lambda_\Omega = 6\frac{\mu\,\Omega}{p_S}\left(\frac{R}{c}\right)^2 = 7,\ \Lambda_\omega = 12\frac{\mu\,\omega}{p_S}\left(\frac{R}{c}\right)^2 = 14$

Without a shaft installed in the bearing, measurements of the air supply pressure (p_S) and ensuing mass flow rate ($\dot{M}_{max}$) deliver an estimation of the permeability coefficient (κ) [20]. From Equation (11),

$$\kappa = \frac{\dot{M}_{max}}{\pi\,D\,L}\underset{c\,\to\infty}{}\,\frac{2\,t_p\,\mu\,R_g\,T}{(p_S{}^2 - p_a{}^2)}. \tag{33}$$

In the measurements, the supply pressure (p_S) increases from 2.4 bar to 6.5 bar (absolute), and an accurate turbine type meter records the flow. Figure 3 shows $\dot{M}_{max}$ vs. $(p_S{}^2 - p_a{}^2)$, and the line fitting the data has a correlation factor $R^2 = 1.0$. From the experimental data, $\kappa = 8.2 \times 10^{-16}\ \mathrm{m}^2$, which is typical for the product.

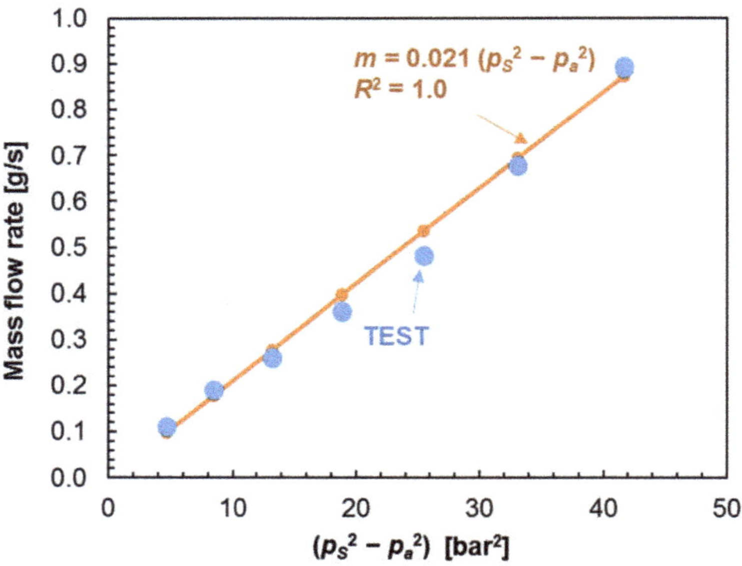

Mass flow rate [g/s]

$(p_S^2 - p_a^2)$ [bar²]

Figure 3. PGB measured mass flow rate $\dot{M}_{max}$ vs. $(p_S^2 - p_a^2)$ and a line fit (no shaft installed).

7. PGB Mass Flow Rate, Peak Pressure, and Aerostatic Stiffness vs. Clearance

From Equation (9), the pressure at the bearing mid-plane ($z = 0$) is $p_{(z=0)} = \sqrt{p_S^2 + (p_a^2 - p_S^2)\left[\cosh\left(\gamma\frac{L}{D}\right)\right]^{-1}}$. For the PGB with geometry listed in Table 1 and $e = 0$, Figure 4 depicts the mass flow rate ($\dot{M}$), $p_{(z=0)}$ and the aerostatic stiffness K_S (= $K_{XXs} = K_{YYs}$) vs. clearance for the PGB supplied with air at $p_S = 6$ bar. Note the equivalent clearance for the porous layer $c_\kappa = (12 \, \kappa \, t_p)^{1/3} \sim 3$ μm.

The graphs include results from the exact solution and the FE computational model (1872 elements = 72 circumferential × 26 axial). Note that shaft speed has no effect on both parameters; see Reynolds Equation (7). The FE model flowrate is almost identical to the exact solution (difference less than 1%). For small clearances ($c < 0.040$ mm), $\dot{M}$ decreases quickly, $p_{(z=0)} \to p_S$, K_S reaches a maximum (184 MN/m) at $c = 0.010$ mm, and then sharply decreases as $c \to 0$. On the other hand, for $c > 0.06$ mm, $\dot{M} \to \dot{M}_{max}= 0.73$ g/s, $\underset{c\to\infty}{}$ $p_{(z=0)} \to p_a$, and $K_S \to 0$, i.e., a complete loss of load carrying capacity. Note $c = 0.010$ mm ($D/c = 7610$) ~ $3.3 \, c_\kappa$, though appearing to be tight (too small bearing clearances increase cost and make installation difficult), is quite appropriate for a gas bearing. Hence, the selection of the PGB clearance is most important to both keep a low flow rate while also providing enough load capacity and centering ability.

Having established the validity of the analytical solution versus a numerical solution, Figure 5 depicts $p_{(z = 0)}/p_a$, mass flow rate, and the (centering) aerostatic stiffness (K_S) vs. clearance (c) for operation with supply pressure (p_S) = 2, 4, 6, and 8 bar. The clearance ranges from a thin $c = 0.002$ mm ($D/c = 38,105$) to $c = 0.1$ mm ($D/c = 762$). Clearly, the film peak pressure, mass flow, and aerostatic stiffness increase with p_S. As noted earlier, too large clearances ($c > 0.040$ mm $\to D/c < 1,900$) produce no film pressure ($p \to p_a$); hence, the PGB leaks too much ($\to \dot{M}_{max}$) and is devoid of load capacity ($K_S \to 0$). For a very tight $\underset{c\to\infty}{}$ clearance ($c < 0.005$ mm), the mid-plane pressure $p_{z=0} \to p_S$, the flow ($\dot{M}$) is small, and the direct stiffness is along a decreasing path. Independent of the supply pressure, the peak aerostatic stiffness appears at the same (narrow) clearance, $c \sim 0.010$ mm = $3.3 \, c_\kappa$. At this c, $p_{z = 0}/p_a = 0.92$ and $\dot{M}/\dot{M}_{max} = 0.37$.

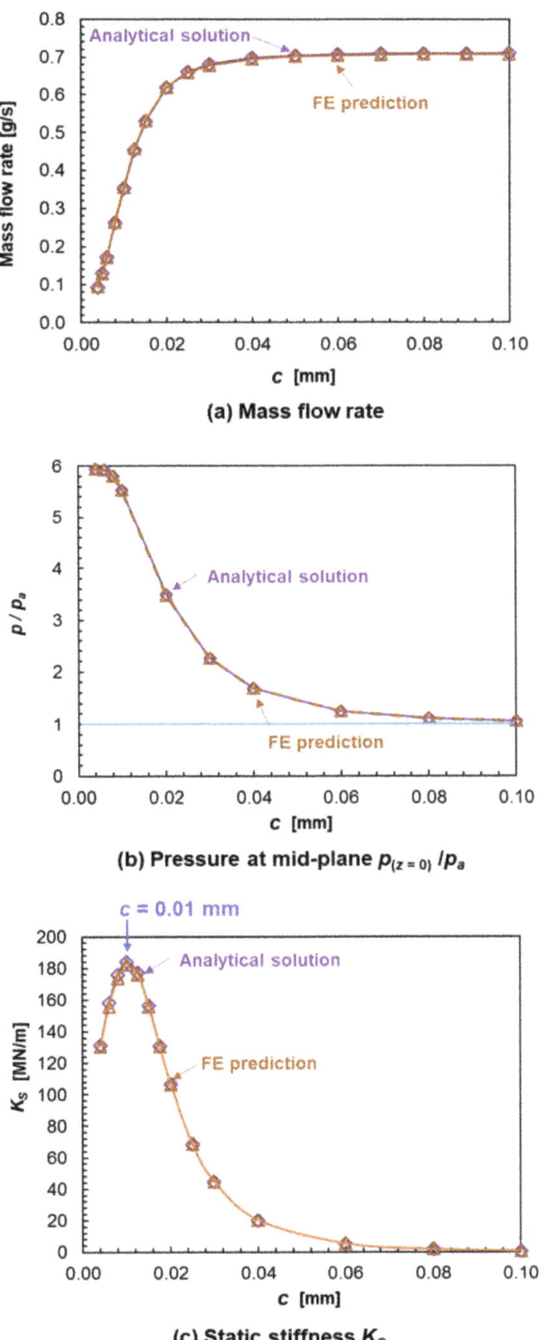

Figure 4. PGB mass flow, pressure at bearing mid-plane, and aerostatic stiffness vs. clearance (c); (**a**) Mass flow rate; (**b**) Pressure at mid-plane $p_{(z\,=\,0)}/p_a$; (**c**) Static stiffness K_S. Analytical and FE predictions. Air supply pressure $p_S = 6$ bar. Aerostatic operation at $e = 0$.

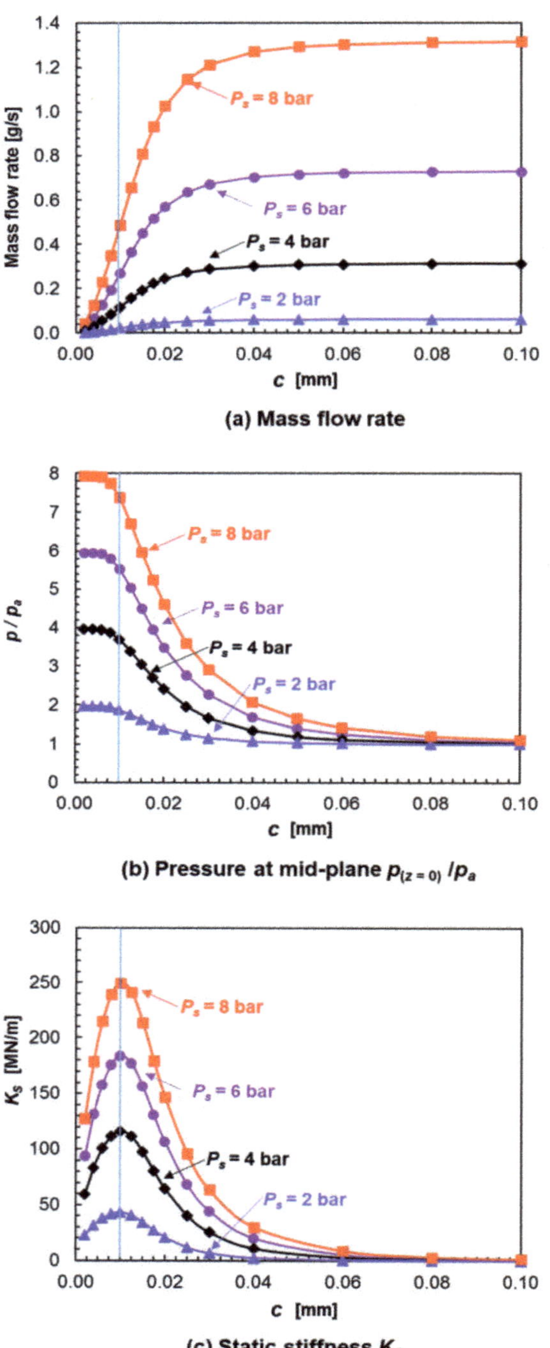

Figure 5. PGB static pressure at mid-plane, mass flow rate, and aerostatic stiffness vs. clearance (c). Air supply pressure p_s = 2, 4, 6, and 8 bar; (**a**) Mass flow rate; (**b**) Pressure at mid-plane $p_{(z=0)}/p_a$; (**c**) Static stiffness K_S. Aerostatic operation ($\Omega = 0$) and null eccentricity ($e = 0$). Vertical lines denote clearance c = 0.01 mm for largest stiffness.

Although not shown for brevity, a dimensionless aerostatic stiffness scaled as $K = (K_S\, c/W_*)$ clusters the data in Figure 5c. At $c = 0.010$ mm, $K \sim 0.6$ for all p_S. Being of order (1), the parametrization of K allows one to quickly estimate the (maximum) centering stiffness of any PGB having the same (κ/t_p). The finding is nearly the same as that reported by Mori et al. [11,12] in 1968, $K \sim 0.7$.

The PGB product specification sheet [27] recommends a PGB with clearance $c = 0.010$ mm and, for operation with pressure supply $p_S = 5.15$ bar [60 psig], quotes a static stiffness $K_S = 159$ MN/m and flow of 13.2 L per minute (LPM) at standard conditions. For the same conditions, the current analysis predicts $K_S = 155.5$ MN/m and a flow rate equaling to 9.86 LPM. The agreement in stiffness is remarkable, while the difference in flows indicates differing clearances. For example, for $c = 0.012$ mm (~20% larger), the analysis predicts 13.2 LPM flow and $K_S = 149$ MN/m. Since the porous layer is affixed to the bearing housing using an adhesive, the manufacturer [27] specifies a maximum operating supply pressure of 7.9 bar (100 psig).

8. PGB Force Coefficients vs. Rotor Speed (Synchronous Frequency Condition)

The analysis continues for the PGB with dimensions in Table 1 and for clearance $c^* = 0.010$ mm, the one providing the peak aerostatic stiffness (K_S). The operating speed range covers 0 to 30 krpm, with a mean operating speed (MOS) $\Omega^* = 25$ krpm (417 Hz); hence the shaft surface speed $\Omega R = 120$ m/s. The pressure supply $p_S = 2$ bar $\rightarrow 8$ bar, and the ambient pressure $p_a = 1$ bar. Note the feed flow parameter $\Lambda_\kappa = \gamma^2 = 5.3$; and at the lowest supply pressure and MOS, $\Lambda_{\Omega*} = 6\frac{\mu\,\Omega_*}{p_S}\left(\frac{R^2}{c^2}\right) \sim 21$, thus indicating a moderate aerodynamic effect will assist with the generation of load capacity.

Figure 6 depicts the PGB force coefficients vs. shaft speed (Ω) and four p_S. The coefficients are evaluated at a whirl frequency coinciding with shaft speed $(\omega = \Omega)$, namely a synchronous speed condition. Note that at the centered condition, $K_{XX} = K_{YY}$, $K_{XY} = -K_{YX}$, $C_{XX} = C_{YY}$, and $C_{XY} = -C_{YX}$; see Equation (22). Effective force coefficients are

$$K_{eff} = K_{XX} + C_{XY}\,\omega, \quad C_{eff} = C_{XX} - K_{XY}/\omega. \tag{34}$$

In general, the magnitude of the PGB force coefficients increases as the supply pressure grows. For hybrid operation (aerostatic plus aerodynamic effects), the direct stiffness (K_{XX}) increases with rotor speed (Ω), whereas the direct damping C_{XX} drops rapidly. At the top speed (30 krpm), K_{XX} is approximately two times larger than at the low speed (aerostatic) condition. On the other hand, C_{XX} reduces to ~1/5 in magnitude at low shaft speed. The cross-coupled stiffness K_{XY} $(<<K_{XX})$ is peculiar, as it reverses sign at a certain shaft speed. The larger p_S is, the higher the speed at which K_{YX} changes from positive to negative. Note $-C_{XY} << C_{XX}$ with a peak magnitude at a certain shaft speed. Since $C_{XY} < 0$, then $K_{eff} < K_{XX}$ for all shaft speeds. At a low shaft speed condition, the effective damping is ~50% of C_{XX}. On the other hand, at the high end of shaft speeds, C_{eff} is slightly larger than C_{XX} since $K_{XY} < 0$.

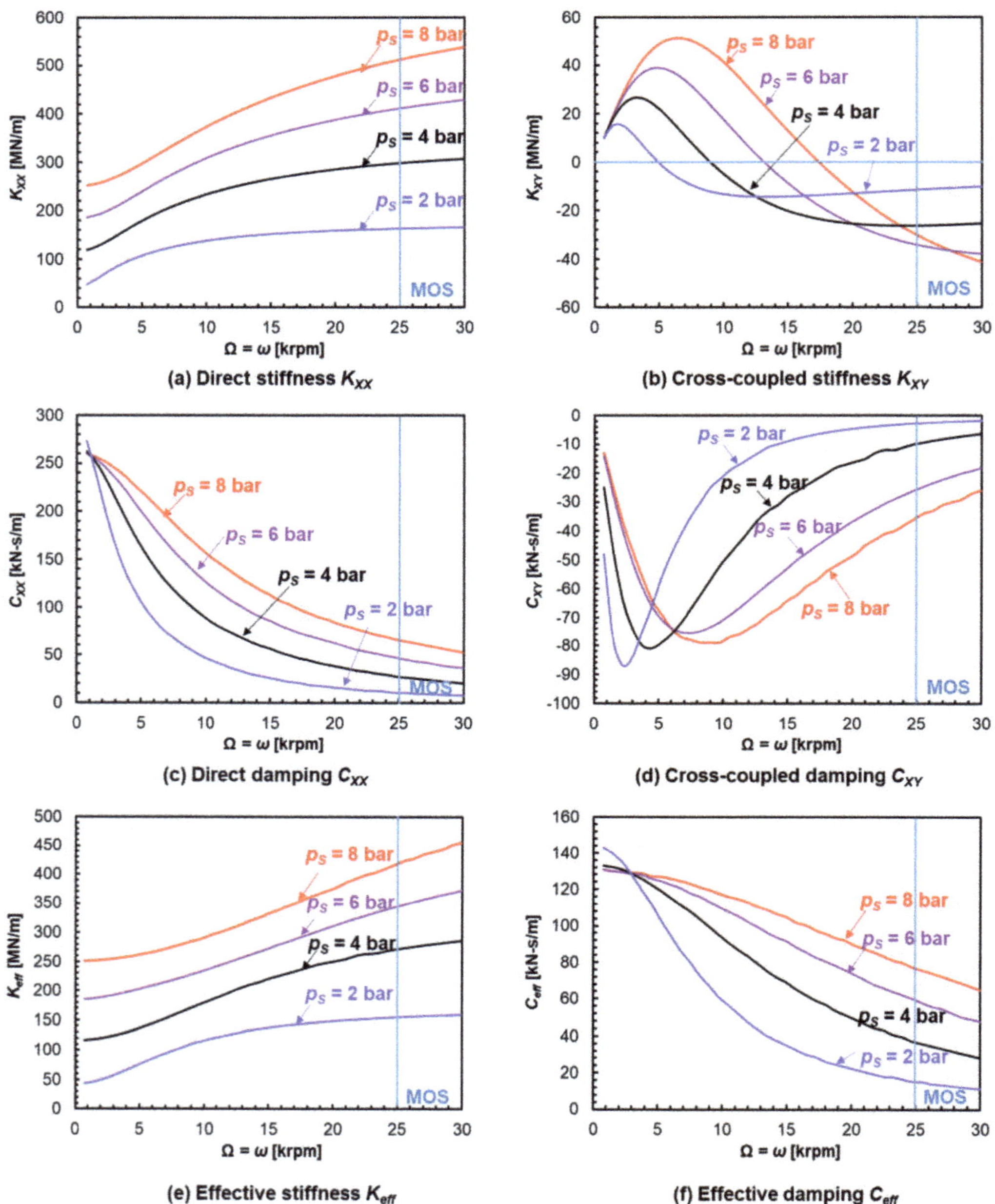

Figure 6. PGB (synchronous frequency) force coefficients vs. shaft speed: (**a**,**b**) stiffnesses (K_{XX}, K_{XY}), (**c**,**d**) damping coefficients (C_{XX}, C_{XY}), (**e**,**f**) effective stiffness (K_{eff}) and damping (C_{eff}) coefficients. Excitation frequency $\omega = \Omega$ (synchronous with speed). Supply pressure p_S = 2 to 8 bar; MOS = 25 krpm.

9. PGB Force Coefficients vs. Excitation Frequency

Figure 7 displays the PGB force coefficients vs. frequency ratio (ω/Ω^*) for the bearing supplied with air at increasing pressures. In general, K_{XX} grows (hardens) as the whirl frequency increases, whereas the direct damping (C_{XX}) quickly drops. The magnitude of all force coefficients increases with an increase in gas supply pressure. In addition, at the cross-over frequency, $\omega = 1/2\,\Omega$, half-frequency whirl, the effective stiffness (K_{eff}) has a dip, and the effective damping turns positive, $C_{eff} > 0$. Hence, the PGB has the identical aerodynamic stability characteristics as a plain journal bearing. The predictions reveal that the cross-coupled stiffness (K_{XY}) reverses sign for $\omega > 1/2\,\Omega$, this frequency increasing as the supply pressure grows. Incidentally, for a low-pressure supply, $p_S < 4$ bar, the direct damping is negative, $C_{XX} < 0$, at low whirl frequencies ($\omega << \Omega^*$). This effect is remarkable and a precursor to a pneumatic hammer; see [13].

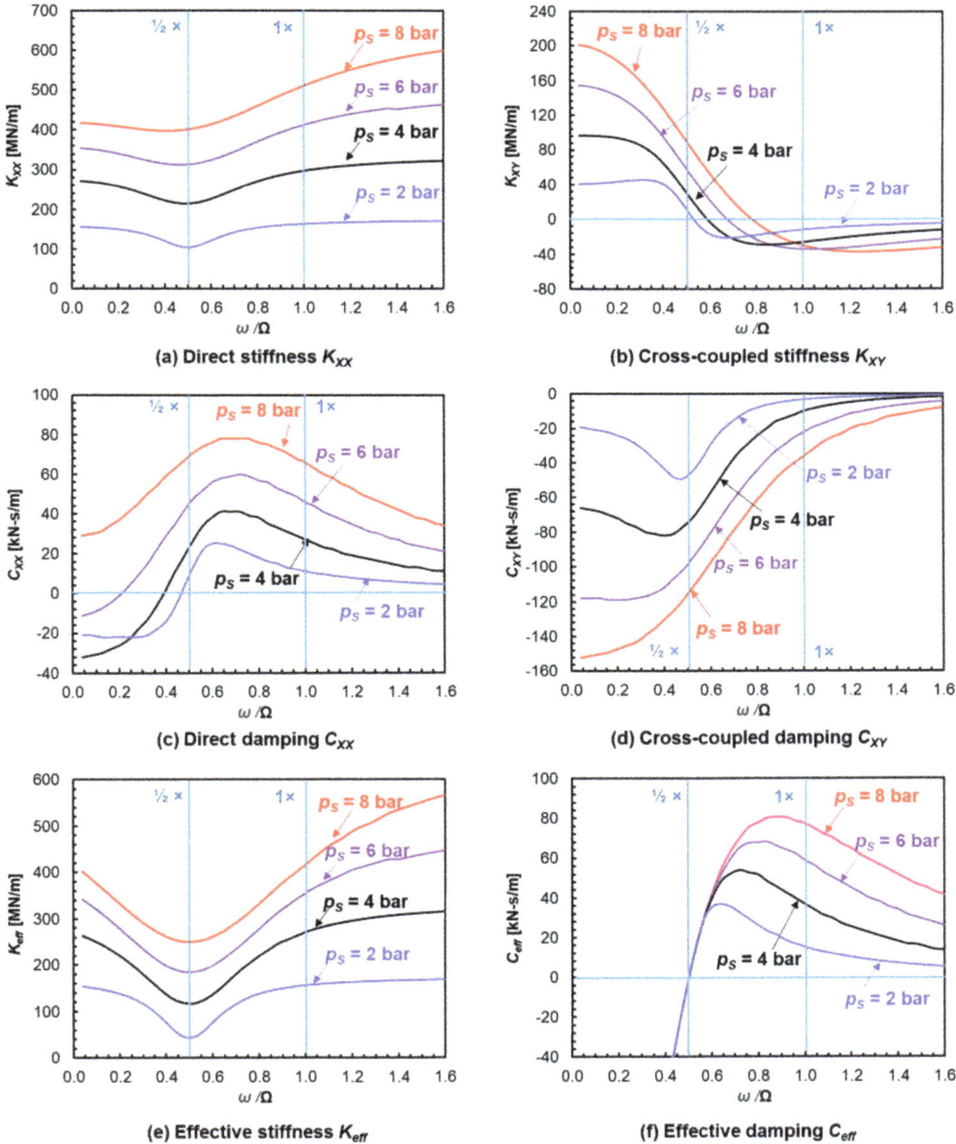

Figure 7. PGB force coefficients vs. frequency ratio (ω/Ω). Mean operating speed $\Omega^* = 25$ krpm (417 Hz); (**a,b**) stiffnesses (K_{XX}, K_{XY}), (**c,d**) damping coefficients (C_{XX}, C_{XY}), (**e,f**) effective stiffness (K_{eff}) and damping (C_{eff}) coefficients. Supply pressure $p_S = 2$ to 8 bar.

In sum, shaft whirl motions at super synchronous motions ($\omega > 1/2\ \Omega$), produce stiffness hardening and a marked reduction in effective damping. On the other hand, SSVs are likely to occur at $1/2$ frequency whirl since $C_{\mathit{eff}} = 0$.

10. Stability of PGB

The stability of a point mass rotor (M_r) supported on one PGB is derived from $\left| \mathbf{H}_{(\omega)} - \omega^2 M_r\, \mathbf{I} \right| = 0$, where $\mathbf{I}$ is the 2×2 identity matrix. Following San Andrés [1], the instability threshold occurs at frequency ω_s defined when the equivalent complex stiffness H_e has its imaginary part Im $(H_e) = 0$ and its real part > zero. The equivalent complex stiffness (H_e) is defined as

$$H_{e_{(\omega)}} = \frac{1}{2}\left(H_{XX_{(\omega)}} + H_{YY_{(\omega)}}\right) - \left[\frac{1}{4}\left(H_{XX_{(\omega)}} - H_{YY_{(\omega)}}\right)^2 + H_{XY_{(\omega)}} H_{YX_{(\omega)}}\right]^{1/2}. \tag{35}$$

Recall that at the centered condition ($e = 0$), $H_{XX} = H_{YY}, H_{XY} = -H_{YX}$. Hence, H_e becomes

$$H_{e_{(\omega)}} = H_{XX_{(\omega)}} - i\, H_{XY_{(\omega)}} = (K_{XX} + \omega C_{XY})_{(\omega)} + i(\omega C_{XX} - K_{XY})_{(\omega)} = K_{\mathit{eff}_{(\omega)}} + i\,\omega C_{\mathit{eff}_{(\omega)}} \tag{36}$$

$$\mathrm{Im}\left(H_{e_{(\omega)}}\right) = 0 \rightarrow C_{\mathit{eff}_{(\omega)}} = 0 \,;\, \mathrm{Re}\left(H_{e_{(\omega)}}\right) > 0 \rightarrow M_{cr} = \frac{K_{\mathit{eff}_{(\omega)}}}{\omega^2}. \tag{37}$$

When $\omega = \frac{1}{2}\Omega \rightarrow C_{\mathit{eff}_{(\omega)}} = 0$, and taking the operating speed as being the threshold speed of instability ($\Omega_T = \Omega^* = 2\omega$), then the largest mass the rotating system can hold is $M_{cr} = \dfrac{K_{\mathit{eff}_{(\frac{1}{2}\Omega_T)}}}{\left(\frac{1}{2}\Omega_T\right)^2}$. Table 2 lists the critical rotor mass for operation at various supply pressures and at MOS = 25 krpm. M_{cr} increases from 26.2 kg to 146.1 kg as $p_S = 2$ bar $\rightarrow$ 8 bar. Note the weight of M_{cr} is much smaller than the PGB load capacity of $\sim(K_\Omega \times c)$. Here $K_\Omega = \sqrt{K_{XX}^2 + K_{XY}^2}$ is a hybrid mode operation stiffness evaluated at the MOS (Ω^*) and a zero whirl frequency ($\omega = 0$).

Table 2. Aerostatic and hybrid stiffnesses for PGB at the centered condition and rotor speed $\Omega = 0$ and 25 krpm. Air supply pressure $p_S = 2, 4, 6,$ and 8 bar. Clearance $c = 0.010$ mm ($\Lambda_\kappa = 5.3$).

Pressure	Aerostatic ($\Omega = 0$)	Hybrid	Ω = 25 krpm		Attitude Angle		Critical Mass
p_S Bar	K_S MN/m	K_{XX} MN/m	K_{XY} MN/m	K_Ω MN/m	β degrees	$K_\Omega K_S$	M_{cr} kg
2 bar	43	156	41	162	14.6	3.76	26.2
4 bar	116	271	96	288	19.6	2.48	68.8
6 bar	184	354	155	387	23.6	2.10	107.7
8 bar	250	418	201	464	25.7	1.85	146.1
	$K_{XY} = 0$						

11. PGB Load Capacity and Attitude Angle

Predictions of the load carrying ability of the PGB for off-centered operation ($e > 0$) follow. Figure 8 depicts the load (W) vs. eccentricity ratio (e/c) for aerostatic operation (0 rpm) and at the mean MOS, $\Omega^* = 25$ krpm, and increasing magnitudes of air supply pressure. The results from the FE computational model (1872 elements = 72 circumferential $\times$ 26 axial), shown with dark symbols, reveal W increases with p_S and is proportional to journal eccentricity (e). The graphs include light-colored lines that represent an approximate load derived from the product of the (exact) static stiffness at $e = 0$ times the journal eccentricity, i.e., $W_{\mathit{approx}} = (K_S \times e)$ when $\Omega = 0$ (left graph), and $W_{\mathit{approx}} = (K_\Omega \times e)$ at the MOS. Table 2 lists the magnitudes of the analytical force coefficients for the aerostatic ($\Omega = 0$) and hybrid (Ω^*) conditions.

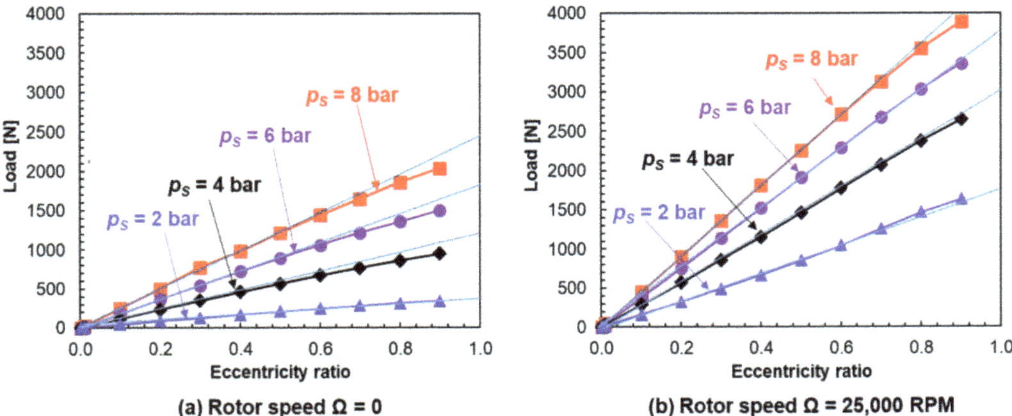

Figure 8. FE model of PGB load (*W*) vs. journal eccentricity ratio (*e/c*) at 0 rpm and 25 krpm shaft speed. Air supply pressure p_S = 2, 4, 6, and 8 bar; (**a**) Load at 0 rpm; (**b**) Load at 25 krpm. PGB clearance *c* = 0.010 mm. (Light colored) straight lines represent approximate load capacity derived from the exact solution at *e* = 0.

Note the remarkable agreement between the analytical solution and the FE numerical solution for operation with journal eccentricity (*e*) as large as 70% of the bearing clearance. In short, the load capacity of the cylindrical PGB is proportional to the journal displacement (*e*). Furthermore, for the hybrid mode operating condition, the attitude angle (*β*) between the eccentricity vector and the applied load vector remains constant. As seen in Table 2, *β* increases from 14.6° to 25.7° as the supply pressure increases from two bar to eight bar.

The analysis and predictions demonstrate that the PGB is a linear mechanical element. Presently, based on the results shown in Figure 8, an estimation for the bearing load capacity equals $\frac{W}{W^*} = \frac{W}{(p_S - p_a)L\,D} \sim \left[0.5 + 0.3\frac{p_a}{p_S}\right](1 + 0.13\Lambda_\Omega)$.

12. An Example of Validation for the Static Performance of a PGB

Li et al. [23] detailed an investigation of the static and dynamic performance of an (in-house constructed) cylindrical PGB. As in the current paper, the authors of [23] built a computational finite difference (FD) model for prediction of performance of simple PGBs and to make pertinent distinctions for operation as purely aerostatic (Ω = 0), aerodynamic ($p_S = p_a$), and hybrid (Ω > 0, $p_S > p_a$). Authors in [6] also detailed a handful of experimental results for the measurement of applied load vs. journal eccentricity for a PGB with two distinct clearances, small (16 μm) and large (31 μm). Table 3 details the geometry, porous layer physical properties, and the operating conditions of the test PGB in [23]. Note that the reference has many clerical errors including dubious captions in several of the figure captions. In addition, the direct force coefficients presented in [23] follow trends not in accordance with the theory, the direct stiffness not agreeing with one derived from the static load ($\omega \rightarrow 0$). The work also gives a cursory review of the cross-coupled force coefficients.

Figure 9 presents comparisons of the current analytical solution and FE model predictions vis-a-vis those in [23], experimental and numerical. As Figure 9a shows, the analytical solution and FE prediction of journal eccentricity (*e*) agrees well with the FD prediction [23] for p_S = 4.7 bar, *c* = 31 μm and three static loads, 5 N to 15 N. For the top load of 15 N, the specific load $W/(LD)$ = 0.1 bar; hence $W/(LD)/(p_S - p_a)$ = 0.029, i.e., a very small load. The model predictions are slightly larger than the experimental results (ref. [23] does not report uncertainty or variability for the measured parameters nor variability for the force coefficients).

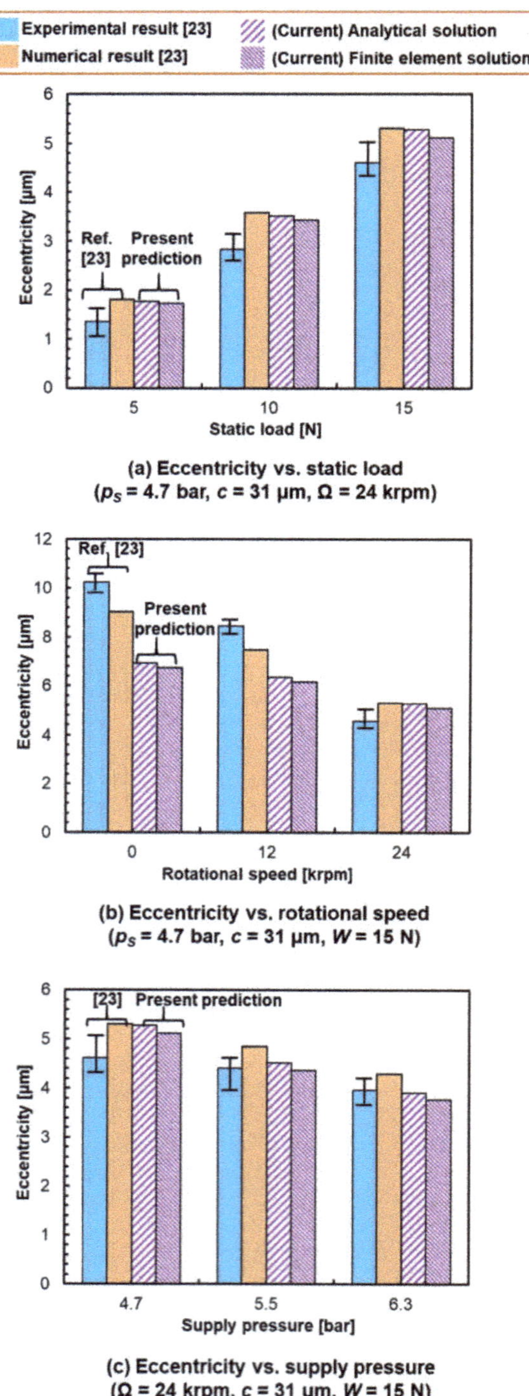

(a) Eccentricity vs. static load
(p_S = 4.7 bar, c = 31 μm, Ω = 24 krpm)

(b) Eccentricity vs. rotational speed
(p_S = 4.7 bar, c = 31 μm, W = 15 N)

(c) Eccentricity vs. supply pressure
(Ω = 24 krpm, c = 31 μm, W = 15 N)

Figure 9. Predictions from current models (analytical and FE) compared to experimental results and predictions reported in [23]. (**a**) Journal eccentricity (*e*) vs. static load (*W*); (**b**) eccentricity (*e*) vs. shaft speed (*Ω*); (**c**) eccentricity (*e*) vs. supply pressure (p_S).

Table 3. Operating conditions and dimensions of a cylindrical PGB reported in Ref. [23].

Bearing Length, L	57 mm
Inner diameter, D	25 mm (*)
Radial clearance, c	16 μm, 31 μm
Porous layer radial thickness, t_p	2.5 mm
permeability coefficient, κ	1.0×10^{-15} m^2
Equivalent clearance porous layer, c_κ	3.1 μm
Supply pressure, p_S	4–6 bar (*)
Exit pressure, p_a	1 bar
Ambient temperature, T	293 K
Air density at (p_a, T), ρ_a	1.2 kg/m^3
viscosity at (p_a, T), μ	18.5×10^{-6} Pa-s
Parameters	For c = 0.031 mm, p_S = 4.7 bar, and Ω = 2513 rad/s (24 krpm)
Feed flow parameter	Λ_κ = 0.025
Speed number	Λ_Ω = 0.097

(*) Assumed since [23] has many clerical mistakes, as confirmed by one of the authors [28]. Ref. [23] also uses gauge pressure to report p_s.

As per the journal eccentricity (e) vs. shaft speed and a fixed load (W = 15 N) (see Figure 9b), the current analytical and FE model predictions agree well with the experimental result at the top speed of 24 krpm. For pure aerostatic operation (Ω = 0), the current model predictions are ~32% smaller than the experimental result. Figure 9c displays eccentricity (e) vs. supply pressure (p_S) for operation at 24 krpm (Λ_Ω = 0.097 at p_S = 4.7 bar). For p_S = 5.5 and 6.3 bar, the current predictions match the experimental values, the difference <2%. For p_S = 4.7 bar, the models produce a slightly larger (e) than the measured one.

Comparisons of predictions of the experimental force coefficients are omitted. As frequency grows, the test coefficients follow differing paths and have lower physical magnitudes than the model predictions, current and those in [23]. Feng [28] attributes the difference to the absence of a surface restrictive layer in the constructed PGB. Otsu et al. [18] explain difference in the performance of PGBs with and without the said restrictive layer. Interestingly enough, the current predictive model as well as that in [23] do not apply to PGBs with restrictive layers.

13. Conclusions

Porous surface gas bearings (PGBs) have come of age to enable high-speed, near friction free rotating machinery with improved reliability and availability. In the last decade (2012 and onward), numerous computational analyses for calculation of PGB forced performance, static and dynamic, have appeared. Most analyses, however, tackled the flow physical equations by computer while ignoring the vast body of archival literature on the subject.

The paper reviews the archival literature on PGBs and reproduces, using modest analytical means, an exact solution to the flow field and forced performance of a cylindrical PGB. This solution has been available since 1979 [5].

Predictions from the analysis serve to validate the results of a finite element computational model. Further comparisons to recent experimental results validate the analysis results and serve to elucidate the effect of external pressure and shaft speed on PGB performance. The learning from this work is as follows:

- For a given external pressure, the supplied flow rate increases quickly as the bearing clearance enlarges and ultimately reaches a flow limit.
- There is a narrow clearance region that ensures the maximum centering stiffness for a PGB. Selecting the appropriate clearance is necessary and rather difficult to

achieve when also considering manufacturing costs and devising procedures for easy installation.

- The load capacity of a PGB under aerostatic conditions is a fraction of the imposed pressure difference and the bearing projected area ($L \times D$). The bearing load is proportional to the static eccentricity.
- Under aerodynamic conditions, i.e., operation with shaft speed, the PGB load capacity still remains proportional to shaft eccentricity and can be much larger than the aerostatic load. That is, shaft speed shear flow effects increase the PGB load capacity.
- For operation as the shaft speed varies from low (start-up) to the mean operating speed (MOS = Ω^*) and above, the PGB bearing shows synchronous excitation ($\omega = \Omega$) force coefficients that increase in magnitude as p_S increases. Most importantly, as the shaft speed increases, K_{eff} increases (hardens) while C_{eff} decreases rapidly.
- For operation at a constant (high) speed, the bearing effective stiffness (K_{eff}) decreases at low whirl frequencies, reaches a dip or minimum at $1/2$ whirl frequency operation ($\omega = 1/2\,\Omega$), and then increases (hardens) as the frequency approaches synchronous speed ($\omega \to \Omega$) and surpasses it. The bearing effective damping coefficient $C_{eff} < 0$ at low frequencies and equals zero at $\omega = 1/2\,\Omega$. For larger ω, $C_{eff} > 0$ and reaches a peak at a certain frequency; the larger the external pressure p_S, the higher the frequency at which C_{eff} is a maximum. For larger frequencies ($\omega >> \Omega$), $C_{eff} \to 0$.
- Note a PGB operating with shaft rotation (hybrid mode) has the same stability restriction as a plain cylindrical hydrodynamic bearing, i.e., a 50% whirl frequency ratio. However, a rigid rotor–PGB system natural frequency is rather large, since the bearing centering stiffness grows with both external pressurization and shaft speed. Hence, the threshold speed of instability, equal to two times the system natural frequency, can be tailored to exceed the system operating speed.

The authors hope other analysts, engineering students in particular, appreciate the effort to obtain the solution, the one that exactly predicts bearing performance and allows quick assessment of trends and selection of physical parameters. In a world ruled by complex numerical models attacking physics with computers, fundamental mathematics does bring meaning to an engineering endeavor as well as personal solace.

Author Contributions: Conceptualization, L.S.A. and A.D.; Formal analysis, L.S.A.; Funding acquisition, L.S.A.; Investigation, J.Y.; Project administration, L.S.A.; Resources, L.S.A. and A.D.; Software, J.Y.; Supervision, L.S.A.; Validation, J.Y.; Visualization, L.S.A. and J.Y.; Writing—original draft, L.S.A.; Writing—review & editing, L.S.A. and J.Y. All authors have read and agreed to the published version of the manuscript.

Funding: The research was funded by Texas A&M University-CONACYT Collaborative Research Program, and Texas A&M University J. Mike Walker '66 Department of Mechanical Engineering 2020 Seed Grant.

Institutional Review Board Statement: Not applicable.

Informed Consent Statement: Not applicable.

Data Availability Statement: Not applicable.

Acknowledgments: Thanks to the Texas A&M University-CONACYT Collaborative Research Program, Texas A&M University J. Mike Walker '66 Department of Mechanical Engineering 2020 Seed Grant, for partial financial support of the research. Thanks to NewWay® Air Bearings for donating the bearing.

Conflicts of Interest: The authors declare no conflict of interest.

Nomenclature

c	Bearing radial clearance (m)
c_κ	$(12\,\kappa\,t_p)^{1/3}$. Equivalent clearance for layer of porous material (m)
C_{eff}	$C_{eff} = (C_{XX} - K_{XY}/\omega)$. Effective damping coefficient (N-s/m)
C_{XX}, C_{YY}	Direct damping coefficients (N-s/m)
C_{XY}, C_{YX}	Cross-coupled damping coefficients (N-s/m)
D	$2R$. Rotor diameter (m)
e	Journal eccentricity (m)
(e_X, e_Y)	Components of journal eccentricity (m), $e = \sqrt{e_X^2 + e_Y^2}$
(F_X, F_Y)	Bearing reaction force components along X and Y directions (N)
H_e	Equivalent complex stiffness at threshold speed of instability (N/m)
H_{XX}, H_{YY}	Direct complex stiffness coefficients (N/m)
H_{XY}, H_{YX}	Cross coupled complex stiffness coefficients (N/m)
h	Film thickness (m)
K_S	PGB aerostatic (zero frequency) stiffness coefficient (N/m)
K_{eff}	$K_{eff} = K_{XX} + C_{XY} \cdot \omega$. Effective stiffness coefficient (N/m)
K_Ω	$\sqrt{K_{XX}^2 + K_{XY}^2}$. PGB hybrid (zero frequency) stiffness coefficient (N/m)
K_{XX}, K_{YY}	Direct stiffness coefficients (MN/m)
K_{XY}, K_{YX}	Cross-coupled stiffness coefficients (MN/m)
L	Bearing axial length (m)
M_r	Mass of point rotor (kg)
M_{cr}	$M_{cr} = K_{eff}/\varpi^2$. Rotor critical mass (kg)
$\dot{M}$	Mass flow rate through a porous gas bearing (kg/s)
$\dot{M}_{max}$	Maximum mass flow rate for a porous gas bearing as $c \to \infty$ (kg/s)
p	Absolute pressure (bar)
p_S, p_a	Supply and ambient absolute pressures (bar)
R_g	Gas constant (J/(kg K))
T	Supply/ambient temperature (K)
T_o	Drag torque (Nm). Power loss = $(T_o\,\Omega)$
t_p	Porous layer radial thickness (m)
W	Applied load (N)
W^*	$((p_S - p_a)\,L\,D)$. Nominal load for aerostatic operation
$x = R\theta, z$	Coordinate system on bearing surface
X, Y	Cartesian coordinate system
β	Attitude angle (deg)
$\Lambda_\kappa = \gamma^2$	$\frac{3\,\kappa}{t_p c}\left(\frac{D}{c}\right)^2$. PGB feed flow parameter
Λ_Ω	$6\frac{\mu\,\Omega}{p_s}\left(\frac{R}{c}\right)^2$. PGB speed number, $\overline{\Lambda}_\Omega = \Lambda_\Omega\left(\frac{p_s}{p_0}\right)$
Λ_ω	$12\frac{\mu\,\omega}{p_s}\left(\frac{R}{c}\right)^2$. PGB frequency number. $\overline{\Lambda}_\omega = \Lambda_\omega\left(\frac{p_s}{p_0}\right)$
κ	Permeability coefficient for the porous material (m^2)
μ	Gas absolute viscosity (Pa-s)
ρ	$p/(R_g T)$. Gas density (kg/m^3)
θ	Circumferential coordinate (-)
ω	Whirl frequency (rad/s)
ϖ	$=\omega_n = 1/2\,\Omega_T$. Whirl frequency = natural frequency at threshold speed of instability (rad/s)
Ω	Rotor speed (rad/s)
Ω_T	Threshold speed of rotor stability (rad/s)

Abbreviations

FE	Finite element
FD	Finite difference
LPM	Liter per minute
MOS	Mean operating speed
PGB	Porous gas bearing
SSV	Subsynchronous whirl vibration

References

1. San Andrés, L. *Modern Lubrication Theory, Notes 15: Gas Bearings for Oil-Free MTM*; Texas A&M University Digital Libraries: College Station, TX, USA, 2020. Available online: http://oaktrust.library.tamu.edu/handle/1969.1/93197 (accessed on 27 June 2021).
2. San Andrés, L.; Yang, J.; McGowan, R. Measurements of Static and Dynamic Load Performance of a 102 mm Carbon-Graphite Porous Surface Tilting-Pad Gas Journal Bearing. *ASME J. Eng. Gas Turbines Power* **2021**. in print. [CrossRef]
3. New Way Air Bearings. Air Bearing Application and Design Guide. Revision E. January 2006. Available online: https://www.newwayairbearings.com/sites/default/files/new_way_application_and_design_guide_%20Rev_E_2006-01-18.pdf (accessed on 27 June 2021).
4. San Andrés, L.; Yang, J.; Devitt, A. About Tilting Pad Carbon-Graphite Porous Gas Bearings: Measurements in a Rotordynamic Test Rig and Predictions. *Tribol. Trans.* **2021**. in print. [CrossRef]
5. Gargiulo, E., Jr. Porous Wall Gas Lubricated Journal Bearings: Theoretical Investigation. *ASME J. Lubr. Tech.* **1979**, *101*, 458–465. [CrossRef]
6. Sneck, H.; Yen, K. The Externally Pressurized, Porous Wall, Gas Lubricated Journal Bearing. *ASLE Trans.* **1964**, *7*, 288–298. [CrossRef]
7. Sneck, H.; Elwell, R. The Externally Pressurized, Porous Wall, Gas Lubricated Journal Bearing. *ASLE Trans.* **1965**, *8*, 339–345. [CrossRef]
8. Sneck, H.; Yen, K. The Externally Pressurized, Porous Wall, Gas Lubricated Journal Bearing. *ASLE Trans.* **1967**, *10*, 339–347. [CrossRef]
9. Kwan, Y.P.B.; Corbett, J. Porous Aerostatic Bearings—An Updated Review. *Wear* **1998**, *222*, 69–73. [CrossRef]
10. Belforte, G.; Raparelli, T.; Viktorov, V.; Trivella, A. Permeability and Inertial Coefficients of Porous Media for Air Bearing Feeding Systems. *ASME J. Tribol.* **2007**, *129*, 705–711. [CrossRef]
11. Mori, H.; Yabe, H.; Yamakage, H.; Furukawa, J. Theoretical Analysis of Externally Pressurized Porous Journal Gas Bearings (1st Report). *Bull JSME* **1968**, *11*, 527–535. [CrossRef]
12. Mori, H.; Yabe, H.; Yamakage, H. Theoretical Analysis of Externally Pressurized Porous Journal Gas Bearings (2nd Report, Journal Bearing with Solid Sleeves). *Bull. JSME* **1968**, *11*, 1512–1518. [CrossRef]
13. Rao, N.S.; Majumdar, B.C. Dynamic Stiffness and Damping Coefficients of Aerostatic, Porous, Journal Gas Bearings. *J. Mech. Eng. Sci.* **1978**, *20*, 291–296. [CrossRef]
14. Hwang, P. *Pneumatic Hammer. Encyclopedia of Tribology*; Springer: Berlin/Heidelberg, Germany, 2013; pp. 2545–2548.
15. Lund, J.W. Calculation of Stiffness and Damping Properties of Gas Bearings. *ASME J. Lubr. Tech.* **1968**, *90*, 793–804. [CrossRef]
16. Gargiulo, E., Jr. Porous Wall Gas Lubricated Journal Bearings: Experimental Investigation. *ASME J. Lubr. Tech.* **1979**, *101*, 466–473. [CrossRef]
17. Miyatake, M.; Yoshimoto, S.; Sato, J. Whirling Instability of a Rotor Supported by Aerostatic Porous Journal Bearings with a Surface-Restricted Layer. *Proc. Inst. Mech. Eng. Part J J. Eng. Tribol.* **2006**, *220*, 95–103. [CrossRef]
18. Otsu, Y.; Miyatake, M.; Yoshimoto, S. Dynamic Characteristics of Aerostatic Porous Journal Bearings with a Surface-Restricted Layer. *ASME J. Tribol.* **2011**, *133*, 011701. [CrossRef]
19. San Andrés, L.; Jeung, S.-H.; Rohmer, M.; Devitt, D. Experimental Assessment of Drag and Rotordynamic Response for A Porous Type Gas Bearing. In Proceedings of the Extended Abstract, STLE Annual Meeting, Dallas, TX, USA, 17–21 May 2015; pp. 17–21.
20. San Andrés, L.; Cable, A.T.; Zheng, Y.; De Santiago, O.; Devitt, D. Assessment of Porous Type Gas Bearings: Measurements of Bearing Performance and Rotor Vibrations. In Proceedings of the ASME Turbo Expo 2016: Turbine Technical Conference and Exposition, Seoul, Korea, 13–17 June 2016; ASME Paper GT2016-57876.
21. Feng, K.; Wu, Y.; Liu, W.; Zhao, X.; Li, W. Theoretical Investigation on Porous Tilting Pad Bearings Considering Tilting Pad Motion and Porous Material Restriction. *Precis. Eng.* **2018**, *53*, 26–37. [CrossRef]
22. Liu, W.; Feng, K.; Huo, Y.; Guo, Z. Measurements of the Rotordynamic Response of a Rotor Supported on Porous Type Gas Bearing. *ASME J. Eng. Gas Turbines Power* **2018**, *140*, 102501. [CrossRef]
23. Li, W.; Wang, S.; Zhao, K.; Feng, K.; Hou, W. Numerical and Experimental Investigation on the Performance of Hybrid Porous Gas Journal Bearings. *Lubr. Sci.* **2020**, *33*, 60–68. [CrossRef]
24. Bohle, M. Numerical Investigation of the Flow in Hydrostatic Journal Bearings with Porous Materials. In Proceedings of the ASME 2018 5th Joint US-European Fluids Engineering Division Summer Meeting, Montreal, QC, Canada, 15–20 July 2018; paper No. FEDSM 2018-83437.
25. Lawrence, T.; Kemple, M. Numerical Method for Studying Bearing Gap Pressure Wave Development and Subsequent Performance Mapping of Externally Pressurized Gas Journal Bearings. In Proceedings of the ASME-JSME-KSME 8th Joint Fluids Engineering Conference, San Francisco, CA, USA, 28 July–1 August 2019; Paper No. AJKFluids 2019-4732.
26. San Andrés, L. Hybrid Flexure Pivot-Tilting Pad Gas Bearings: Analysis and Experimental Validation. *ASME J. Tribol.* **2006**, *128*, 551–558. [CrossRef]
27. NewWay®Air Bearings. Specification Sheet for Porous MediaTM Air Bushing, 3.00 Inch Inside Diameter. Part No. 3307501. Available online: https://www.newwayairbearings.com/catalog/product/air-bushings-english-3-00-inch-id/ (accessed on 27 June 2021).
28. Feng, K. State Key Laboratory of Advanced Design and Manufacturing for Vehicle Body, Hunan University, Changsha, China. Personal communication, 2021.

Article

Parameter Sensitivity Analysis on Dynamic Coefficients of Partial Arc Annular-Thrust Aerostatic Porous Journal Bearings

Pyung Hwang [1,*], Polina Khan [2] and Seok-Won Kang [1,*]

1   Department of Automotive Engineering, Yeungnam University, Gyeongsan 38541, Korea
2   Melentiev Energy Systems Institute of Siberian Branch of Russian Academy of Sciences, 664033 Irkutsk, Russia; polinakhan@isem.irk.ru
*   Correspondence: phwang@yu.ac.kr (P.H.); swkang@yu.ac.kr (S.-W.K.)

Abstract: Aerostatic bearings are widely used in high-precision devices. Partial arc annular-thrust aerostatic porous journal bearings are a prominent type of aerostatic bearings, which carry both radial and axial loads and provide high load-carrying capacity, low air consumption, and relatively low cost. Spindle shaft tilting is a resource-demanding challenge in numerical modeling because it involves a 3D air flow. In this study, the air flow problem was solved using a COMSOL software, and the dynamic coefficients for tilting degrees of freedom were obtained using finite differences. The obtained results exhibit significant coupling between the tilting motion in the x-and y-directions: cross-coupled coefficients can achieve 20% of the direct coefficient for stiffness and 50% for damping. In addition, a nonlinear behavior can be expected, because the tilting motion within 3°, tilting velocities within 0.0012°/s, and relative eccentricity of 0.2 have effects as large as 20% for direct stiffness and 100% for cross-coupled stiffness and damping. All dynamic coefficients were fitted with a polynomial of eccentricity, tilting, and tilting velocities in two directions, with a total of six parameters. The resulting fitting coefficient tables can be employed for the fast dynamic simulation of the rotor shaft carried on the proposed bearing type.

Keywords: aerostatic journal bearings; porous bearing; annular-thrust bearing; numerical simulation; finite element method; dynamic coefficients; eccentricity; tilting; tilting velocity; partial arc

Citation: Hwang, P.; Khan, P.; Kang, S.-W. Parameter Sensitivity Analysis on Dynamic Coefficients of Partial Arc Annular-Thrust Aerostatic Porous Journal Bearings. *Appl. Sci.* **2021**, *11*, 10791. https://doi.org/10.3390/app112210791

Academic Editors: Terenziano Raparelli, Federico Colombo, Andrea Trivella and Luigi Lentini

Received: 30 September 2021
Accepted: 11 November 2021
Published: 15 November 2021

Publisher's Note: MDPI stays neutral with regard to jurisdictional claims in published maps and institutional affiliations.

1. Introduction

Porous aerostatic bearings are widely used in machinery, owing to their relatively high load-carrying capacity and low air consumption [1–3]. Similar to other types of aerostatic bearings, owing to their high accuracy and zero contamination, they are suitable for high-precision devices. In addition, the friction drag of a porous aerostatic bearing is low [4,5], its motion accuracy is higher than that of a conventional orifice-type aerostatic bearing [5,6], and its load thresholds at high rotation speeds are the highest among other types of aerostatic bearings [5,7]. The production of porous materials with desirable permeabilities is relatively complicated. However, the amount of porous material required can be reduced via design optimization [8].

The optimal governing equations required for porous air bearing analysis remain under discussion. Zhong et al. [9] experimentally determined Ergun's equation coefficients for the pressure drop in sintered metal porous media for air bearings. Belforte et al. [10] experimentally verified Forchheimer's law for porous air bearing applications. Zhong et al. [11] verified the accuracy of Forchheimer's law for air flow through sintered metal porous media and Darcy's law, under slight pressure drops. Recently, it has been reported that a deep learning-based approach can be employed for high-precision and low-cost analysis of the 3D flow state inside a porous material [12].

Various numerical methods and engineering packages have been adopted for air pressure calculations in porous bearings. Huang et al. [13] analyzed the pressure in a porous conveyor air bearing using the FLUENT software. Cui et al. [14] used FLUENT software to numerically analyze the effect of manufacturing errors on the load-carrying capacity and stiffness of an annular-thrust porous aerostatic bearing. In addition, the deformation of porous thrust bearings was numerically investigated based on the fluid–solid interaction method by Wang et al. [15]. Van Ostayen et al. [16] used the COMSOL software for active aerostatic bearing analysis. Hwang and Khan [17] conducted 3D finite element analysis, while Plante et al. [18] proposed the shed 1D generalized flow theory. Otsu et al. [19] and Hosokawa et al. [8] used the FDM.

Several papers [19–22] deal with the numerical calculation of stiffness and damping coefficients corresponding to eccentricity or parallel shaft displacement (Figure 1a). However, insufficient data are available on stiffness and damping for shaft tilting (Figure 1b) in porous air bushing. Cui et al. [14,23] analyzed the non-flatness effect of the air-bearing surface on the stiffness and damping of an aerostatic porous bearing, which is computationally similar to modeling the tilting.

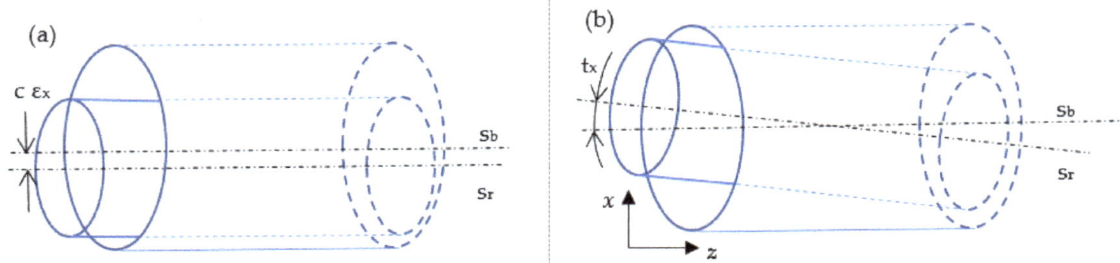

Figure 1. Two spindle motion modes: (**a**) eccentricity; (**b**) tilting. s_b and s_r denote the bearing and rotor axes, respectively.

The rotor in ultraprecision devices is usually supported by journal and thrust bearings, which share a common air supply system but separate porous pads [14,23]. Annular-thrust bearings share the same porous pad carrying loads in both the annular and thrust directions (Figure 2). This dual role decreases the manufacturing cost while triggering flow coupling [23–27]. A partial arc bearing has a higher load-carrying capacity but lower stiffness than a full-arc bearing [27].

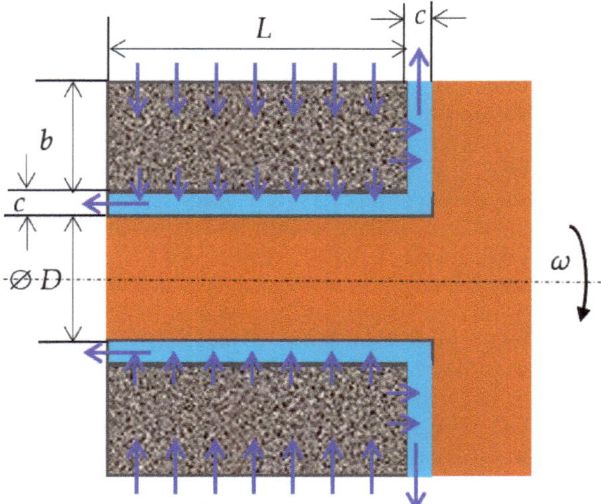

Figure 2. Annular-thrust porous aerostatic bearing.

The numerical modeling of the rotor motion, including the effect of shaft tilting, as well as the modeling of a partial arc annular-thrust porous aerostatic bearing is substantially time-consuming, because it requires a 3D air flow solution. Dynamic modeling can be performed in a less resource-consuming manner if the full bearing model is replaced with stiffness and damping coefficients. The nonlinear dynamic motion can be modeled, provided the influences of eccentricity, tilting, and velocities on the dynamic coefficients are included. In this study, the air flow in the porous pad and air lubrication film between the porous pad and solid surface of the shaft were numerically modeled using the COMSOL Multiphysics® software. The pressure distribution and total forces and moments were calculated for low eccentricity, low tilting angles, and tilting speeds. The stiffness and damping coefficients were determined using finite differences.

2. Materials and Methods

The porous air bearing was modeled using the COMSOL Multiphysics software, "Darcy's law", and "Thin-Film Flow, Shell" modules. Here, the static solution is sufficient in determining the response of the displacement and velocity of the spindle. In addition, the porous bushing is a hollow cylinder parallel to the z-axis with a cut sector α. Figure 3 illustrates the different boundary conditions.

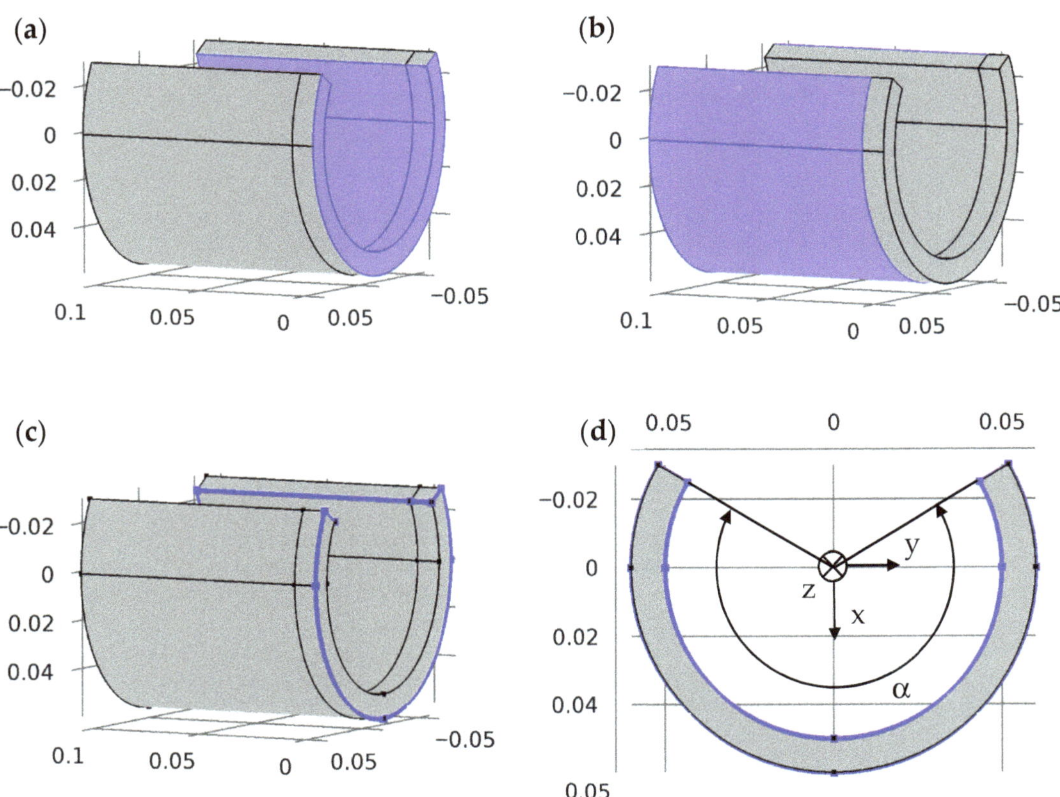

Figure 3. Porous pad boundaries: (**a**) airbearing surface carrying both radial and axial loads; (**b**) air supply surface; (**c**) outflow to atmospheric pressure; (**d**) outflow to atmospheric pressure in *xy*-plane with axes and bearing partial arc.

The pressure in the porous pad is modeled with the "Darcy's law" module:

$$-\nabla\left(\rho\frac{\kappa}{\mu}\nabla p\right) = 0 \tag{1}$$

where p, ρ, κ, and μ denote the air pressure, air density, permeability coefficient, and dynamic viscosity of air, respectively. ∇ is the gradient operator in three-dimensional space. The pressure at the air-bearing surface (see Figure 3a) is modeled with the "thin-film flow, shell" module:

$$\nabla_t\left(\rho h\left(\frac{1}{2}v_t - \frac{h^3}{12\mu}\nabla_t p_f\right)\right) = -\rho\frac{\partial}{\partial t}h + Q \tag{2}$$

where p_f, h, v_t, and Q represent the air film pressure, air film thickness, tangential spindle surface velocity, and air flux, respectively. ∇_t is the gradient operator in the two-dimensional space of the air film surface and ∇ is the partial derivative operator. At the interface between the porous pad and the air-bearing surface, the pressure and air flux must be continuous. In the "Darcy's law" module, the "Pressure" boundary condition is adopted at the surfaces that open to the air lubrication film, which is expressed as:

$$p = p_f. \tag{3}$$

In the "Thin-Film Flow, Shell", the continuity of the air flux is expressed by admittance Y in the "Perforation" interface, which is expressed as:

$$Y = \frac{Q\cdot n}{\rho\, p_f} \tag{4}$$

where $n = (n_x, n_y, n_z)$ is the air-bearing surface normal vector. The air film thickness in the "Thin-Film Flow, Shell" module depends on the spindle eccentricity $(\varepsilon_x, \varepsilon_y)$, tilting (t_x, t_y), and tilting velocities $(\dot{t}_x, \dot{t}_y)$ (refer to Figure 1):

$$h = c + u\cdot n - u\cdot\nabla_t h \tag{5}$$

$$\frac{\partial}{\partial t}h = v\cdot n - v\cdot\nabla_t h \tag{6}$$

$$u = \left(c\varepsilon_x + t_x(z - L/2); c\varepsilon_y + t_y(z - L/2); -t_x x - t_y y\right) \tag{7}$$

$$v = \left(-y\omega + \dot{t}_x(z - L/2); x\omega + \dot{t}_y(z - L/2); -\dot{t}_x x - \dot{t}_y y\right) \tag{8}$$

where u is the spindle surface displacement, and v is the spindle surface velocity.

The isothermal compressibility of the air was modeled using the user-defined density in both modules. Here, the density is related to the air pressure in the "Variables" block as follows:

$$\rho = \frac{\rho_0\, p}{p_0}, \tag{9}$$

where the subscript 0 refers to the atmospheric conditions, and $p_0 = 1$ bar. The air supply is modeled by another "Pressure" interface in the "Darcy's law" module with a constant pressure of p_s (see Figure 3b). The outflow at the air-bearing edges is given by the "Border" interface in the "Thin-Film Flow, Shell" module with a fixed pressure of p_0 (see Figure 3c). The other boundaries were sealed.

After the static problem was solved and the air pressure was calculated, the moments in the x and y tilt directions were calculated, using the "Surface Integration" in the "Derived Values" block, according to the following expressions:

$$M_x = \iint p(n_x(z - L/2) - n_z x)dx'dy' \tag{10}$$

$$M_y = \iint p\left(n_y(z - L/2) - n_z y\right) dx'dy' \tag{11}$$

where x' and y' are the local coordinates of the air film surface.

Stiffness and damping for tilting degrees of freedom were calculated by using finite differences:

$$K_{tx,tx} = -\left(M_x\left(\varepsilon_x, \varepsilon_y, t_x + \Delta t_x, t_y, \dot{t}_x, \dot{t}_y\right) - M_x\left(\varepsilon_x, \varepsilon_y, t_x, t_y, \dot{t}_x, \dot{t}_y\right)\right)/\Delta t_x \tag{12}$$

$$K_{tx,ty} = -\left(M_x\left(\varepsilon_x, \varepsilon_y, t_x, t_y + \Delta t_y, \dot{t}_x, \dot{t}_y\right) - M_x\left(\varepsilon_x, \varepsilon_y, t_x, t_y, \dot{t}_x, \dot{t}_y\right)\right)/\Delta t_y \tag{13}$$

$$K_{ty,tx} = -\left(M_y\left(\varepsilon_x, \varepsilon_y, t_x + \Delta t_x, t_y, \dot{t}_x, \dot{t}_y\right) - M_y\left(\varepsilon_x, \varepsilon_y, t_x, t_y, \dot{t}_x, \dot{t}_y\right)\right)/\Delta t_x \tag{14}$$

$$K_{ty,ty} = -\left(M_y\left(\varepsilon_x, \varepsilon_y, t_x, t_y + \Delta t_y, \dot{t}_x, \dot{t}_y\right) - M_y\left(\varepsilon_x, \varepsilon_y, t_x, t_y, \dot{t}_x, \dot{t}_y\right)\right)/\Delta t_y \tag{15}$$

$$D_{tx,tx} = -\left(M_x\left(\varepsilon_x, \varepsilon_y, t_x, t_y, \dot{t}_x + \Delta \dot{t}_x, \dot{t}_y\right) - M_x\left(\varepsilon_x, \varepsilon_y, t_x, t_y, \dot{t}_x, \dot{t}_y\right)\right)/\Delta \dot{t}_x \tag{16}$$

$$D_{tx,ty} = -\left(M_x\left(\varepsilon_x, \varepsilon_y, t_x, t_y, \dot{t}_x, \dot{t}_y + \Delta \dot{t}_y\right) - M_x\left(\varepsilon_x, \varepsilon_y, t_x, t_y, \dot{t}_x, \dot{t}_y\right)\right)/\Delta \dot{t}_y \tag{17}$$

$$D_{ty,tx} = -\left(M_y\left(\varepsilon_x, \varepsilon_y, t_x, t_y, \dot{t}_x + \Delta \dot{t}_x, \dot{t}_y\right) - M_y\left(\varepsilon_x, \varepsilon_y, t_x, t_y, \dot{t}_x, \dot{t}_y\right)\right)/\Delta \dot{t}_x \tag{18}$$

$$D_{ty,ty} = -\left(M_y\left(\varepsilon_x, \varepsilon_y, t_x, t_y, \dot{t}_x, \dot{t}_y + \Delta \dot{t}_y\right) - M_y\left(\varepsilon_x, \varepsilon_y, t_x, t_y, \dot{t}_x, \dot{t}_y\right)\right)/\Delta \dot{t}_y \tag{19}$$

3. Results

The airbearing moments, tilting stiffness, and damping were calculated for $\varepsilon_x = 0.1$, 0.2, 0.3 , $\varepsilon_y = 0.0$, 0.1, $t_x = -0.003°$, $-0.00275°$, ..., $0.003°$, and $t_y = 0.0°$, $0.00025°$,..., $0.003°$, respectively; with $p_s = 8$ bar, bearing length $L = 0.11$ m, shaft diameter $D = 0.1$ m, porous cylinder thickness $b = 0.01$ m, clearance $c = 10$ μm, rotation speed = 1000 rpm, bearing arc $\alpha = 240°$, porosity = 0.4, and permeability = 8.33×10^{-17} m². The film variation due to tilting in the given range does not exceed 57% for simultaneous change of t_x and t_y. Together with 30% variation due to eccentricity, it gives a minimal film thickness of 13% of clearance, or 1.3 μm. The air-bearing pressure, forces, and moments were calculated for all combinations of parameters t_x, t_y, ε_x, and ε_y, and their combinations with the following pairs of tilting velocity values: $\left(\dot{t}_x, \dot{t}_y\right) \in \left\{(0,0), \left(\Delta \dot{t}_x, 0\right), \left(2\Delta \dot{t}_x, 0\right), \left(0, \Delta \dot{t}_y\right), \left(0, 2\Delta \dot{t}_y\right)\right\}$. The total number of cases was 13000. The steps in the differences (12–19) were $\Delta t_x = \Delta t_y = 0.00025°$, and $\Delta \dot{t}_x = \Delta \dot{t}_y = 0.00573°/s$. The maximum and average absolute errors of the finite differences are presented in Table 1.

Table 1. Finite difference errors.

	Stiffness (kN·m/rad)				Damping (kN·m·s/rad)			
	$K_{tx,tx}$	$K_{tx,ty}$	$K_{ty,tx}$	$K_{ty,ty}$	$D_{tx,tx}$	$D_{tx,ty}$	$D_{ty,tx}$	$D_{ty,ty}$
max	0.124	0.066	0.089	0.120	0.011	0.008	0.003	0.003
average	0.022	0.006	0.010	0.016	0.003	0.002	0.0004	0.0003

3.1. Preliminary Analysis

As illustrated in Figure 4, the direct stiffness coefficient $K_{tx,tx}$ has values in the range of 80–140 kN·m/rad. Within the range under consideration, $K_{tx,tx}$ depends linearly on ε_x and nonlinearly on t_x, t_y, $\dot{t}_x$, $\dot{t}_y$. The parameters t_x, $\dot{t}_x$, and ε_x exhibit significant effects (more than 10%). $K_{tx,tx}$ is almost insensitive to t_x at low negative tilting angles; however, it decreases linearly with t_x at positive and high negative tilting angles (see Figure 4a). The effect of the x-tilting velocity $\dot{t}_x$ is the diminishing of $K_{tx,tx}$, which is more significant at negative tilting angles. $K_{tx,tx}$ increases with decreasing t_y, increasing $\dot{t}_y$ and ε_x, and is insensible to ε_y (see Figure 4b,c).

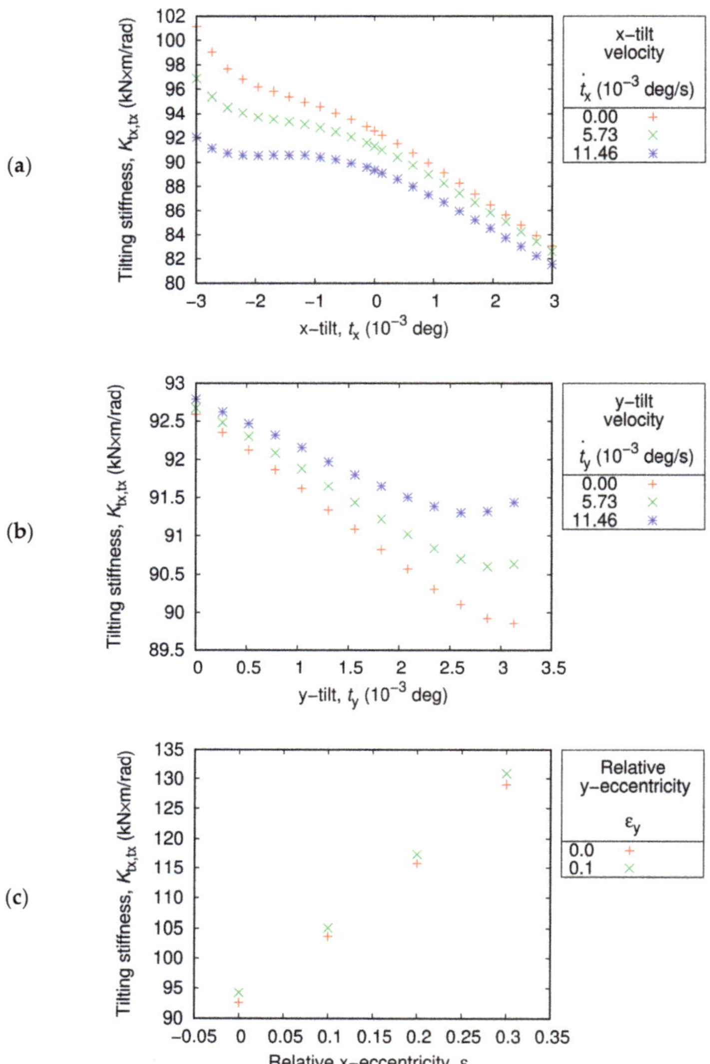

Figure 4. Parameter sensitivity of the direct stiffness coefficient $K_{tx,tx}$: (**a**) Effects of the tilting angle and tilting velocity in the x-direction; (**b**) Effects of the tilting angle and tilting velocity in the y-direction; (**c**) Effects of relative eccentricities. All unreferenced parameters have zero values.

The cross-coupled stiffness coefficient $K_{tx,ty}$ is approximately 10 times smaller than the direct stiffness coefficient $K_{tx,tx}$, and lies in the range of 6–15 kN·m/rad (see Figure 5). It is highly sensitive to all parameters, except for $\dot{t}_x$. $K_{tx,ty}$ increases with decreasing t_x and $\dot{t}_y$, and increasing t_y (see Figure 5a,b). Eccentricity ε_x can increase or decrease $K_{tx,ty}$, depending on ε_y (see Figure 5c).

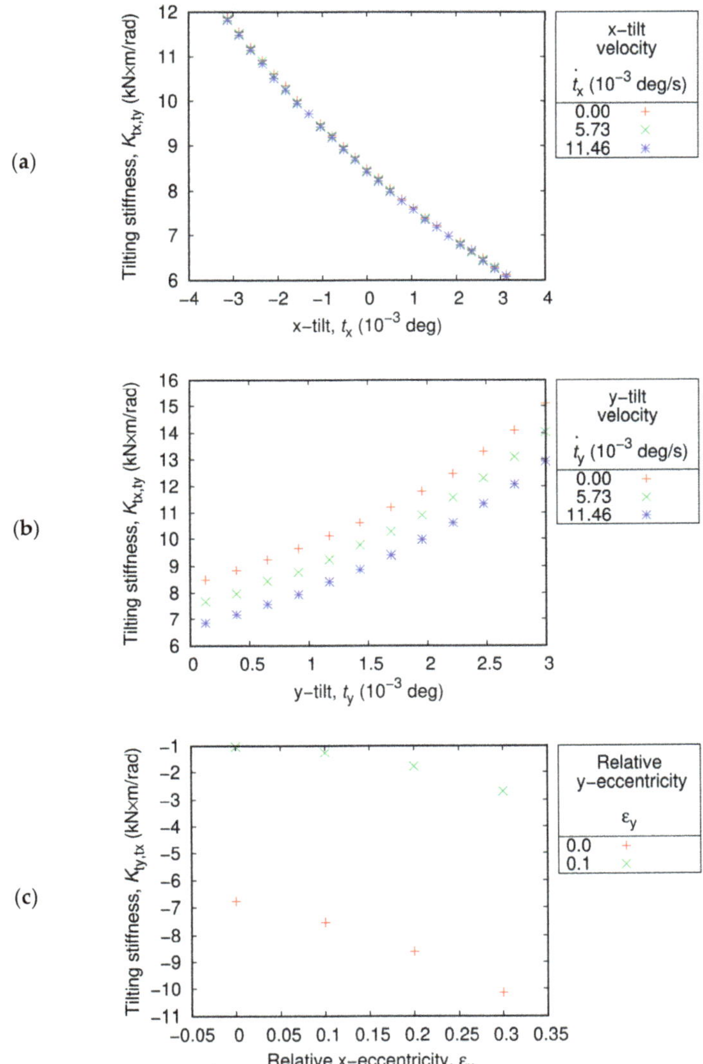

Figure 5. Parameter sensitivity of the cross-coupled stiffness coefficient $K_{tx,ty}$: (**a**) Effects of tilting angle and tilting velocity in the x-direction; (**b**) Effects of the tilting angle and tilting velocity in the y-direction; (**c**) Effects of relative eccentricities. All unreferenced parameters have zero values.

$K_{ty,tx}$, which is another cross-coupled stiffness coefficient, is of the same order of magnitude; however, it is negative, and falls in the range of -18–0 kN·m/rad (refer to Figure 6). It is not sensitive to $\dot{t}_x$, and the $\dot{t}_y$ effect on it is also negligible (see Figure 6a,b). The magnitude of $K_{ty,tx}$ is higher at lower t_x and ε_y and at higher t_y and ε_x.

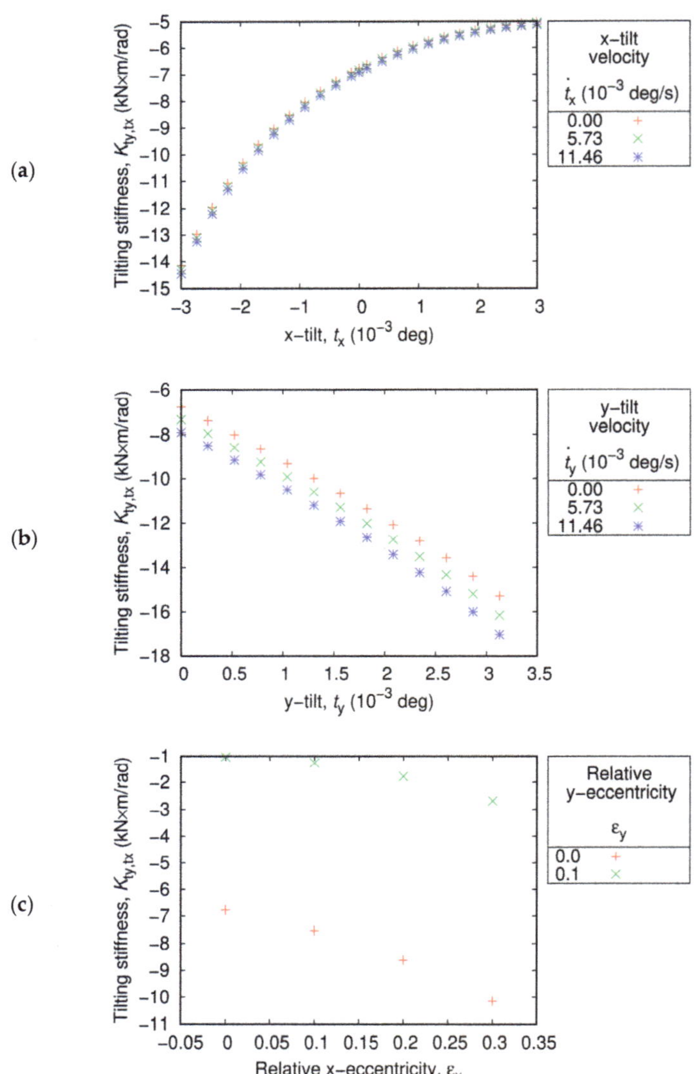

Figure 6. Parameter sensitivity of the cross-coupled stiffness coefficient $K_{ty,tx}$: (**a**) Effects of the tilting angle and tilting velocity in the x-direction; (**b**) Effects of the tilting angle and tilting velocity in the y-direction; (**c**) Effects of relative eccentricities. All unreferenced parameters have zero values.

The direct stiffness coefficient $K_{ty,ty}$ is higher than that of $K_{tx,tx}$. $K_{ty,ty}$ falls in the range of 110–140 kN·m/rad (refer to Figure 7). This is triggered by the cut arc, which increases force and decreases the stiffness in the x-direction [27]. The y-tilting angle t_y exerts a strong nonlinear effect on $K_{ty,ty}$. The stiffness coefficient $K_{ty,ty}$ is almost constant for a low positive t_y, and then increases with increasing t_y. In contrast to $K_{tx,tx}$, which is primarily sensitive to "its own" parameter t_x and significantly less sensitive to t_y. $K_{ty,ty}$ is equally sensitive to both t_x and t_y. In addition, it is equally sensitive to both higher ε_x and ε_y, with a negligible $\dot{t}_y$ effect, and no $\dot{t}_x$ effect. Furthermore, it is higher for lower t_x and higher t_y, ε_x, and ε_y.

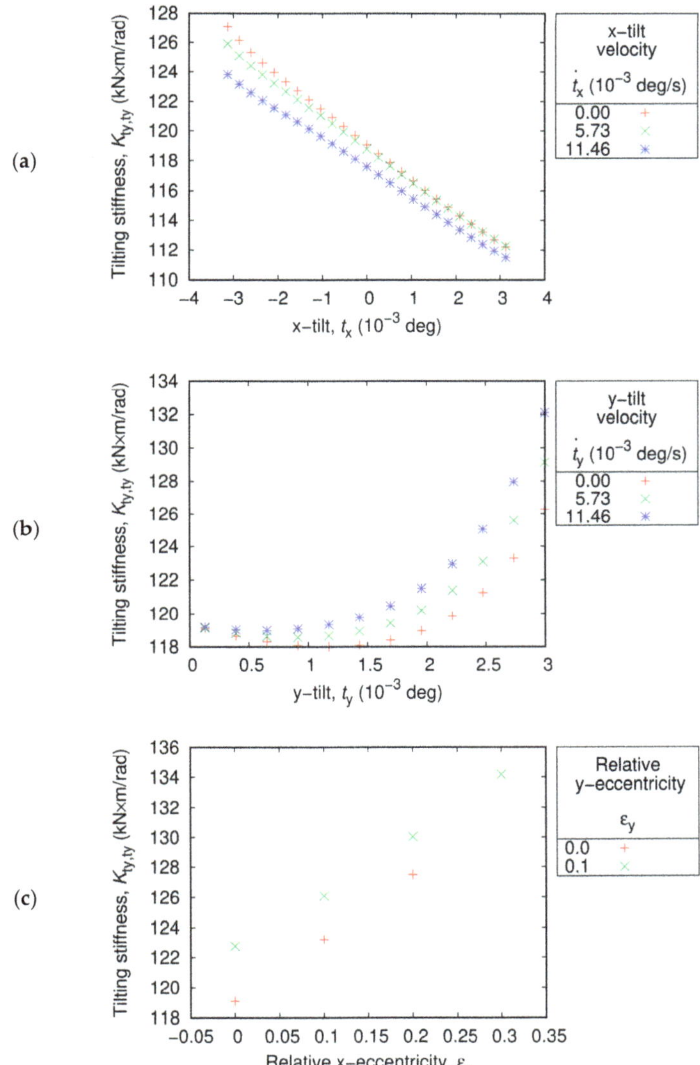

Figure 7. Parameter sensitivity of the direct stiffness coefficient $K_{ty,ty}$: (**a**) Effects of the tilting angle and tilting velocity in the x-direction; (**b**) Effects of the tilting angle and tilting velocity in the y-direction; (**c**) Effects of relative eccentricities. All unreferenced parameters have zero values.

Because the simultaneous variation of tilting velocities was not studied, the effect of $\dot{t}_x$ was solely studied for $D_{tx,tx}$ and $D_{ty,tx}$, while the effect of $\dot{t}_y$ was solely studied for $D_{tx,ty}$ and $D_{ty,ty}$. Instead of the missing parameter, the coupling between t_x and t_y is comprehensively presented in Figures 8–11.

The direct damping coefficient $D_{tx,tx}$ is positive and falls in the range 0.2–2.2 kN·m s/rad (refer to Figure 8). It is higher for lower t_x, and higher t_y and higher $\dot{t}_x$ (see Figure 8a,b). The effect of eccentricity is also substantial, nonlinear, and strongly coupled between ε_x and ε_y (see Figure 8c).

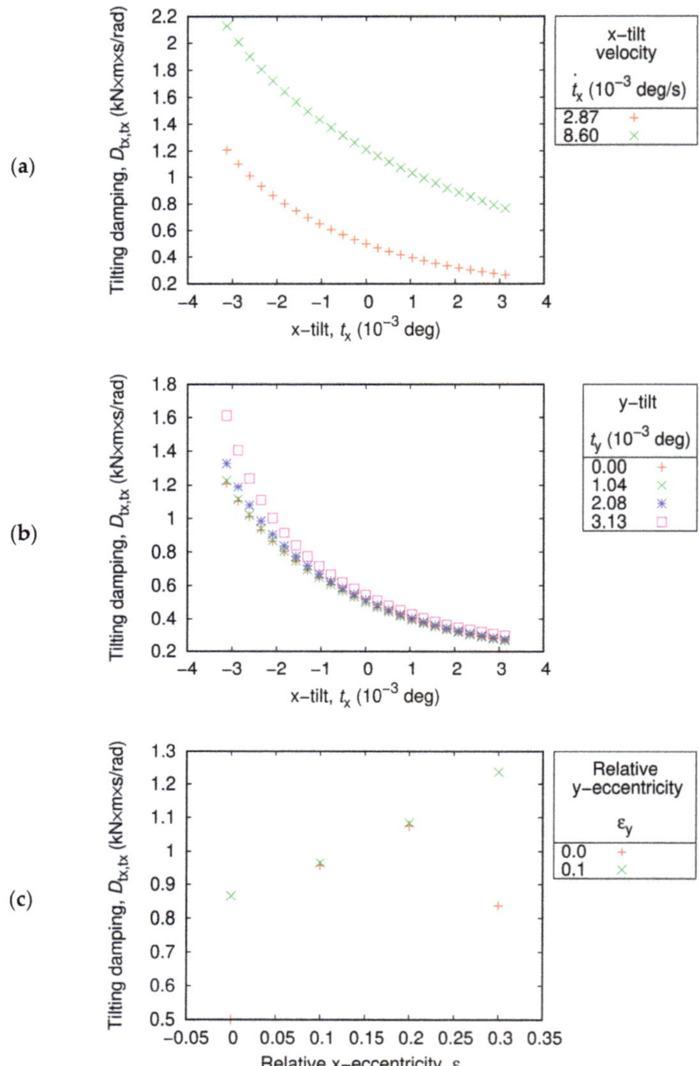

Figure 8. Parameter sensitivity of the direct damping coefficient $D_{tx,tx}$: (**a**) Effects of the tilting angle and tilting velocity in the x-direction; (**b**) Effects of the x and y tilting angles; (**c**) Effects of the relative eccentricities. All unreferenced parameters have zero values.

The cross-coupled damping coefficient $D_{tx,ty}$ does not exceed 1 kN m s/rad in magnitude, being mostly negative (see Figure 9). $D_{tx,ty}$ rapidly decreases with decreasing t_x or increasing t_y in the negative t_x range, while in the positive t_x range, it is much less sensible both to t_x and t_y (see Figure 9a). $D_{tx,ty}$ decreases with increasing t_y and is almost insensible to $\dot{t}_y$ (see Figure 9b). The dependence of $D_{tx,ty}$ on eccentricity parameters is highly nonlinear. $D_{tx,ty}$ can take relatively high positive values for zero ε_x. and positive ε_y, but in most of the range, the sensitivity to ε_x and ε_y is low (see Figure 9c).

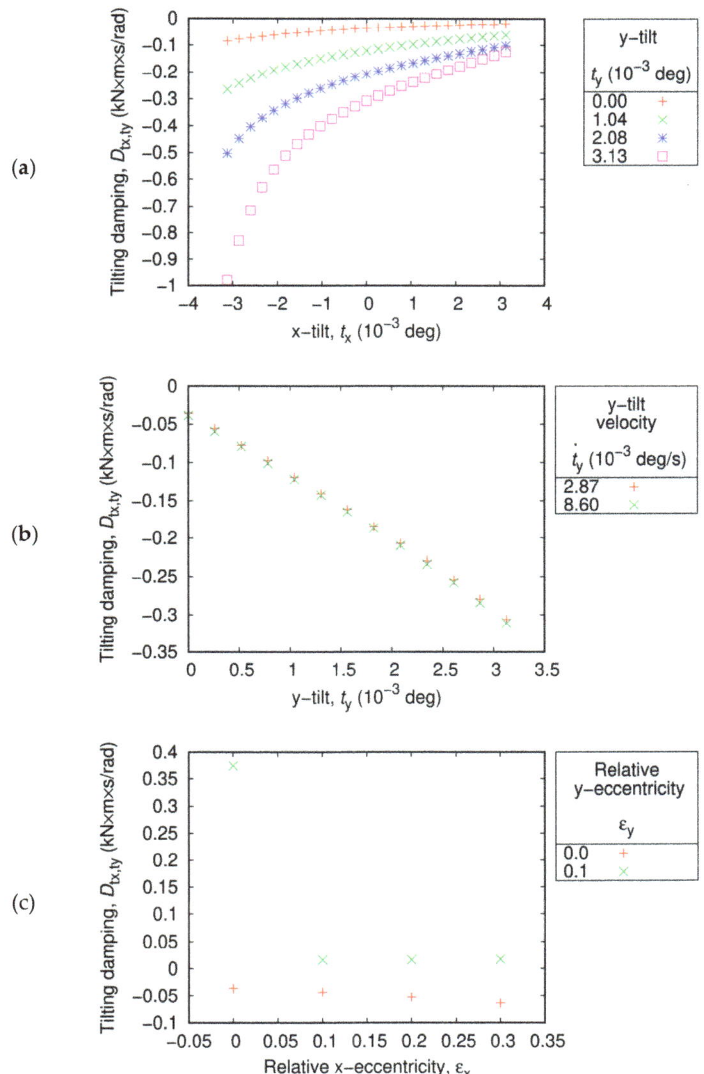

Figure 9. Parameter sensitivity of the cross-coupled damping coefficient $D_{tx,ty}$: (**a**) Effects of x and y tilting angles; (**b**) Effects of the tilting angle and tilting velocity in the y-direction; (**c**) Effects of relative eccentricities. All unreferenced parameters have zero values.

The cross-coupled damping coefficient $D_{ty,tx}$ falls in the range -0.6–0.13 kN m s/rad (see Figure 10). It is negative for positive t_y and negative t_x, otherwise it is positive. Similar to $D_{tx,ty}$, $D_{ty,tx}$ rapidly decreases with decreasing t_x or increasing t_y in the negative t_x range, while in the positive t_x range, it is much less sensible both to t_x and t_y. However, for $t_x > 0.002\,°$, $D_{ty,tx}$ increases with increasing t_y (see Figure 10b), while $D_{tx,ty}$ decreases with increasing t_y for any t_x (see Figure 9a). For zero t_y, $D_{ty,tx}$ decreases with increasing t_x and decreasing $\dot{t}_x$. Similar to all previously considered damping coefficients, the dependence of $D_{ty,tx}$ on the eccentricity parameters is highly nonlinear.

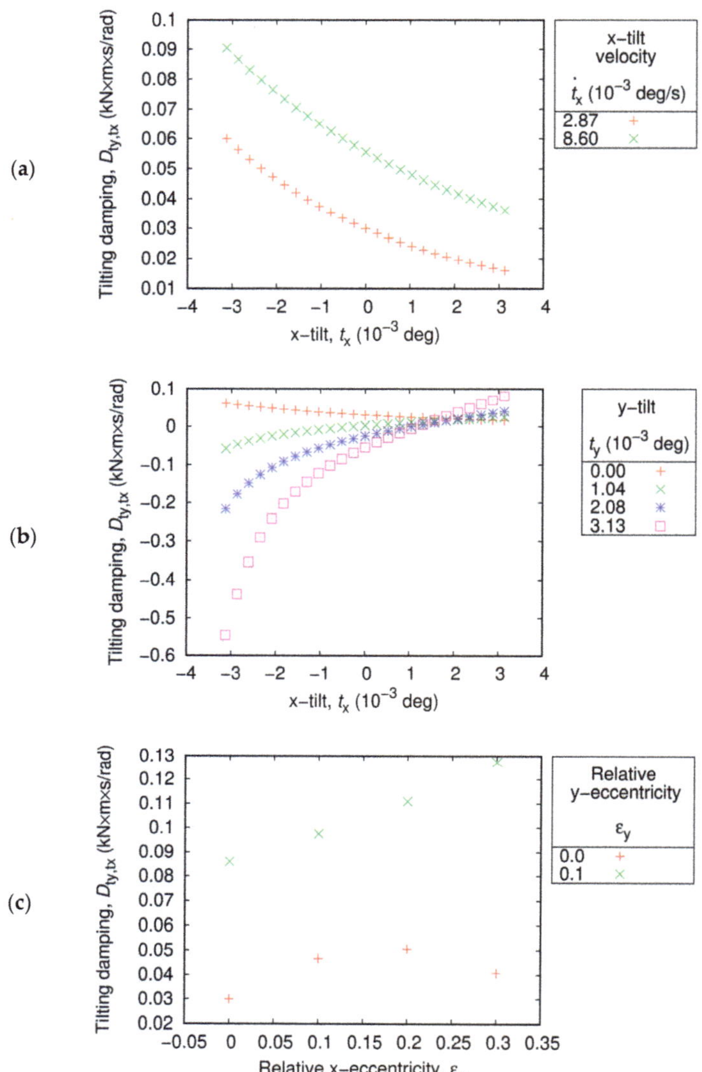

Figure 10. Parameter sensitivity of the cross-coupled damping coefficient $D_{ty,tx}$: (**a**) Effects of the tilting angle and tilting velocity in the x-direction; (**b**) Effects of x and y tilting angles; (**c**) Effects of relative eccentricities. All unreferenced parameters have zero values.

In contrast with all other damping coefficients, the direct damping coefficient $D_{ty,ty}$ does not significantly depend on ε_x and ε_y (refer to Figure 11). It falls in the range 0.6–2.0 kN·m s/rad. It is higher for lower t_x, and higher t_y, similar to $D_{tx,tx}$. $D_{ty,ty}$ is almost insensible to the y-tilt velocity $\dot{t}_y$. It slightly increases with increasing ε_x and increasing ε_y.

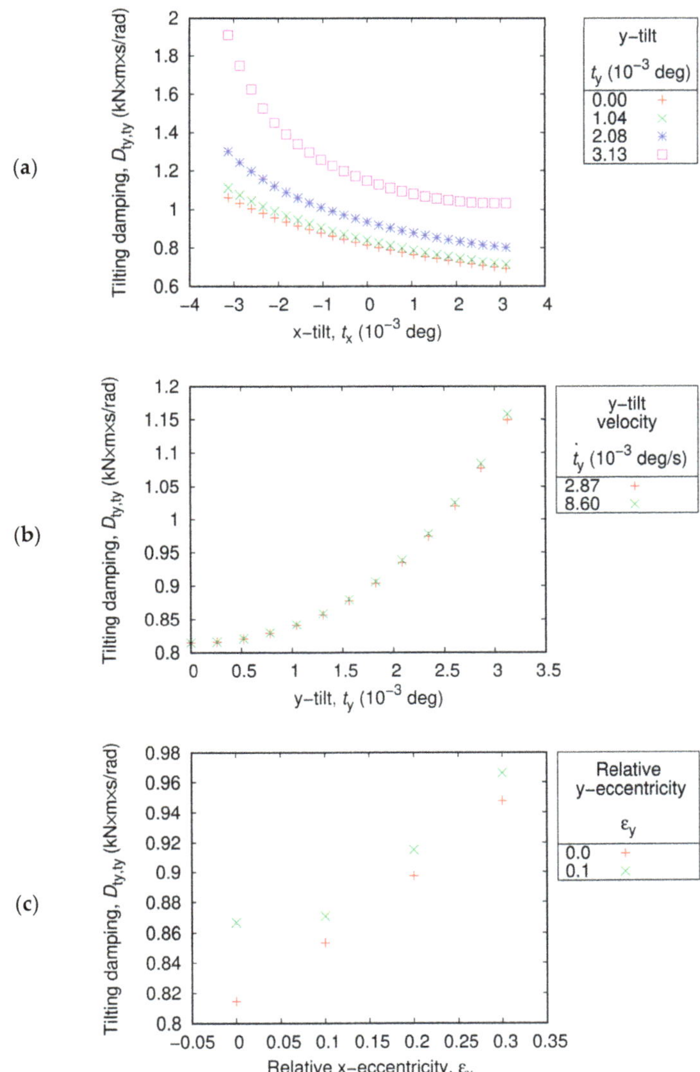

Figure 11. Parameter sensitivity of the direct damping coefficient $D_{ty,ty}$: (**a**) Effects of x and y tilting angles; (**b**) Effects of the tilting angle and tilting velocity in the y-direction; (**c**) Effects of relative eccentricities. All unreferenced parameters have zero values.

In summary, the direct stiffness and damping coefficients were positive. The cross-coupled stiffness coefficients are approximately 10 times smaller than the direct stiffness coefficients, and the cross-coupled damping coefficients are approximately 2–3 times smaller than the direct damping coefficients. Hence, a stable tilting motion was expected. However, information on the rotor inertia and the bearing at the other end is required to make precise conclusions on the dynamic behavior. Moreover, the results above exhibit a strong coupling between the eccentricity and tilting motion, including the nonlinearity of the moments relative to the position parameters. This can make the rotor motion severely complicated.

3.2. Regression

Although it is possible to obtain the direct solution of the time-dependent problem in the model described above, the process is significantly time consuming because the air flow problem must be solved at every step. Conversely, the air-bearing forces and moments can be calculated using the interpolated stiffness and damping coefficients. In this study, each dynamic coefficient for each combination of ε_x, ε_y, $\dot{t}_x$, and $\dot{t}_y$ was first fitted with a quadratic polynomial of t_x and t_y, and then each coefficient of the polynomial was fitted with a linear function of ε_x, ε_y, $\dot{t}_x$, and $\dot{t}_y$. The accuracy of the fitting is expressed in the form of determination coefficients, as presented in Table 2. The accuracy is relatively good, although it can be improved for $\widetilde{K}_{tx,ty}$, $\widetilde{D}_{tx,tx}$, $\widetilde{D}_{tx,ty}$, and $\widetilde{D}_{ty,tx}$ by providing more points and increasing the fitting order in ε_x and ε_y degrees of freedom.

Table 2. Coefficients of determination for dimensionless dynamic coefficient fitting.

	Stiffness (kN·m/rad)				Damping (kN·m·s/rad)			
	$\widetilde{K}_{tx,tx}$	$\widetilde{K}_{tx,ty}$	$\widetilde{K}_{ty,tx}$	$\widetilde{K}_{ty,ty}$	$\widetilde{D}_{tx,tx}$	$\widetilde{D}_{tx,ty}$	$\widetilde{D}_{ty,tx}$	$\widetilde{D}_{ty,ty}$
R^2	0.99	0.77	0.97	0.96	0.89	0.83	0.91	0.96

The fitting results are presented in Tables 3–10. An example of the table interpretation is given below for $K_{tx,tx}$:

$$\widetilde{K}_{tx,tx} = \left(1.09533 + 1.49129\varepsilon_x + 0.17719\varepsilon_y - 0.02161\dot{T}_x\right) +$$
$$\left(-0.30745 - 3.94666\varepsilon_x + 0.10915\dot{T}_x + 0.00478\dot{T}_y\right)T_x +$$
$$\left(-0.15003 + 0.09324\varepsilon_x - 1.16823\varepsilon_y + 0.01582\dot{T}_x + 0.0114\dot{T}_y\right)T_y +$$
$$\left(-0.3232 + 2.45267\varepsilon_x + 2.42515\varepsilon_y + 0.16265\dot{T}_x - 0.28434\dot{T}_y\right)T_xT_y +$$
$$\left(7.99445\varepsilon_x - 0.26854\varepsilon_y - 0.35236\dot{T}_x + 0.14364\dot{T}_y\right)T_x^2 +$$
$$\left(0.34256 + 0.41188\varepsilon_x + 1.09915\varepsilon_y - 0.13117\dot{T}_x + 0.17535\dot{T}_y\right)T_y^2$$

Table 3. Fitting results for $\widetilde{K}_{tx,tx}$.

	1	ε_x	ε_y	$\dot{T}_x$	$\dot{T}_y$
1	1.09533	1.49129	0.17719	−0.02161	0
T_x	−0.30745	−3.94666	0	0.10915	0.00478
T_y	−0.15003	0.09324	−1.16823	0.01582	0.0114
T_xT_y	−0.3232	2.45267	2.42515	0.16265	−0.28434
T_x^2	0	7.99445	−0.26854	−0.35236	0.14364
T_y^2	0.34256	0.41188	1.09915	−0.13117	0.17535

Table 4. Fitting results for $\widetilde{K}_{tx,ty}$.

	1	ε_x	ε_y	$\dot{T}_x$	$\dot{T}_y$
1	0.10973	0	0.51941	−0.00802	−0.02162
T_x	−0.13722	−0.006	−0.80745	0	−0.02419
T_y	0	−1.68577	−0.42286	0.08025	0.193
T_xT_y	0.22904	3.62763	−2.10164	0	0.74463
T_x^2	−0.1213	1.02097	1.42966	0.06678	−0.15351
T_y^2	0.50093	3.62374	−2.3118	0	−0.34029

Table 5. Fitting results for $\widetilde{K}_{ty,tx}$.

	1	ε_x	ε_y	$\dot{T}_x$	$\dot{T}_y$
1	−0.07381	−0.09673	0.78029	−0.00428	−0.00686
T_x	0.11042	0.41995	−1.26002	0.01449	0.00965
T_y	−0.24779	−1.11043	0.33455	0.02062	0.02685
$T_x T_y$	1.06605	2.90228	0	−0.20887	0.24493
T_x^2	−0.72174	−0.86434	1.50162	0.07247	−0.11828
T_y^2	−0.6127	0.69547	0.93397	0.10771	−0.20109

Table 6. Fitting results for $\widetilde{K}_{ty,ty}$.

	1	ε_x	ε_y	$\dot{T}_x$	$\dot{T}_y$
1	1.42569	0.47905	0.29851	0	0.0155
T_x	−0.30423	−1.10565	0	0.0651	0.12299
T_y	−0.64244	0.25697	−3.21467	0	−0.10432
$T_x T_y$	−1.01131	0	5.08434	0.17214	−0.52047
T_x^2	0.43263	1.33588	0	−0.03328	0.27025
T_y^2	3.53634	1.57983	2.17612	−0.13259	0.69417

Table 7. Fitting results for $\widetilde{D}_{tx,tx}$.

	1	ε_x	ε_y	$\dot{T}_x$	$\dot{T}_y$
1	0.0097	0.00358	0	0.00156	0
T_x	−0.02028	−0.05093	0	−0.00117	0
T_y	−0.00168	−0.00288	−0.01683	0.00023	0
$T_x T_y$	−0.01595	−0.04394	0.09286	0.00158	0
T_x^2	0.03404	0.19438	0	−0.00225	0
t_y^2	0.01708	0.04876	−0.01605	−0.0015	0

Table 8. Fitting results for $\widetilde{D}_{tx,ty}$.

	1	ε_x	ε_y	$\dot{T}_x$	$\dot{T}_y$
1	0.0024	−0.00962	0.01265	0	−0.00171
T_x	−0.00299	0.01579	−0.02654	0	0.00227
T_y	−0.00896	−0.01581	0	0	0.00029
$T_x T_y$	0.03654	0.10701	−0.05103	0	0.00167
T_x^2	−0.00931	−0.05237	0.05414	0	−0.00165
T_y^2	−0.01297	−0.03191	0.06242	0	−0.0018

Table 9. Fitting results for $\widetilde{D}_{ty,tx}$.

	1	ε_x	ε_y	$\dot{T}_x$	$\dot{T}_y$
1	0.00072	0.00114	0.00674	0	0
T_x	−0.00151	−0.00441	−0.01802	0	0
T_y	−0.00833	0	−0.01345	−0.00208	0
$T_x T_y$	0.03325	0.08433	−0.02828	0	0
T_x^2	−0.00695	−0.01593	0.04641	0	0
T_y^2	−0.00795	−0.01552	0.0475	0.00093	0

Table 10. Fitting results for $\widetilde{D}_{ty,ty}$.

	1	ε_x	ε_y	$\dot{T}_x$	$\dot{T}_y$
1	0.00987	0.00426	0.00289	0	−0.00014
T_x	−0.00619	−0.01127	−0.0017	0	0.00026
T_y	−0.00943	0.01182	−0.04927	0	0.00208
$T_x T_y$	−0.01952	−0.05093	0.10336	0	−0.00248
T_x^2	0.01514	0.04059	−0.01941	0	0.0007
T_y^2	0.06967	0.05133	−0.07676	0	0.00259

Here, the dimensionless variables are adopted:

$$\frac{K_{i,j}}{\widetilde{K}_{i,j}} = \frac{p_0 D L^3}{16\,c} = 83,187\,\frac{\text{N m}}{\text{rad}} \tag{20}$$

$$\frac{D_{i,j}}{\widetilde{D}_{i,j}} = \frac{p_0 D L^2}{2\left(0.0001\frac{\text{rad}}{\text{s}}\right)} = 1.51 \times 10^5\,\frac{\text{N m s}}{\text{rad}} \tag{21}$$

$$\frac{t_x}{T_x} = \frac{t_y}{T_y} = \frac{c}{L} = 0.000182\,\text{rad} = 0.0104° \tag{22}$$

$$\frac{\dot{t}_x}{\dot{T}_x} = \frac{\dot{t}_y}{\dot{T}_y} = 0.0001\,\frac{\text{rad}}{\text{s}} = 0.00573°\,\frac{1}{\text{s}} \tag{23}$$

4. Discussion

The tilting stiffness and damping analyzed in this study could be significant for the spindle shaft tilting motion, provided the shaft length is not significantly longer than the bearing length. Otherwise, the shaft tilting will be controlled by the air-bearing force applied at a long distance from the shaft center, while the tilting stiffness and damping will influence the shaft bending. In addition, bending can also significantly influence the air-bearing forces, moments, and dynamic coefficients, because of the convexity and concavity of the bearing surface [14,23]. Based on the obtained results, the tilting motion is independently stable because the direct stiffness and damping coefficients are positive and significantly higher in amplitude than the cross-coupled coefficients. However, the tilting dynamic coefficients were substantially influenced by shaft eccentricity. Hence, the stability of the shaft parallel and tilting motions must be analyzed together. The increased rotor speed would result in a more pronounced hydrodynamic component in the bearing forces, moments, stiffness, and damping components, as well as in more significant x-y coupling, that is typical for the journal bearings.

5. Conclusions

The problem of tilting stiffness and damping analysis is hardly ever addressed in the literature since the fundamental books on rotodynamic and tribology where the theoretical analysis was presented for idealized types of bearings. This ignorance can be explained by the minor role of individual bearing moments, compared to the total shaft-bearing systems moments in most applications. However, the tilting moments exist, and they become more and more important in high-precision machinery. Besides the bearing moments, the tilting motion greatly affects the bearing forces, as well as other types of air film distortions, which are widely considered in recent publications.

The purpose of this study was to predict the behavioral characteristics according to the design and operating parameters of porous aerostatic bearings that can be utilized in applications requiring high-precision positioning, such as semiconductor manufacturing equipment. Most importantly, the highly efficient and reliable numerical analysis approach

Appl. Sci. **2021**, *11*, 10791

proposed in this study can contribute to increasing system stability and reducing costs and risks at the product level prior to the actual manufacturing process.

Author Contributions: Conceptualization, P.H. and P.K.; methodology, P.K.; software, P.K.; validation, P.H. and P.K.; formal analysis, P.K.; investigation, P.H. and P.K.; resources, P.H. and P.K.; data curation, P.H.; writing—original draft preparation. P.K.; writing—review and editing. P.H. and S.-W.K.; visualization, P.K.; supervision, P.H.; project administration, P.H.; funding acquisition, P.H. and S.-W.K. All authors have read and agreed to the published version of the manuscript.

Funding: This work was partly supported by Korea Institute for Advancement of Technology (KIAT) grant funded by the Korea Government (MOTIE) (P0002092, The Competency Development Program for Industry Specialist) and Korea Institute of Energy Technology Evaluation and Planning (KETEP) grant funded by the Korea Government (MOTIE) (20214000000010, Gyeongbuk Regional Wind Energy Cluster Human Resources Development Project) and State Assignment Project (No. FWEU-2021-0005) of the Fundamental Research Program of the Russian Federation 2021–2030.

Informed Consent Statement: Not applicable.

Data Availability Statement: Data are available by request.

Acknowledgments: The research was performed with use of the equipment of High-Temperature Circuit Multi-Access Research Center (Project No. 13.ЦКП.21.0038).

Conflicts of Interest: The authors declare no conflict of interest.

References

1. Akilian, M.; Forest, C.R.; Slocum, A.H.; Trumper, D.L.; Schattenburg, M.L. Thin optic constraint. *Precis. Eng.* **2007**, *31*, 130–138. [CrossRef]
2. New Way Air Bearings. *Air Bearing Application and Design Guide*; IBS Precision Engineering: Eindhoven, The Netherlands, 2006.
3. Khim, G.; Park, C.H. A rotary stage in a vacuum using an air bearing. *Vacuum* **2014**, *105*, 39–45. [CrossRef]
4. Andres, L.S.; Cable, T.A.; Zheng, Y.; De Santiago, O.; Devitt, D. Assessment of Porous Type Gas Bearings: Measurements of Bearing Performance and Rotor Vibrations. In Proceedings of the ASME Turbo Expo 2016: Turbomachinery Technical Conference and Exposition, Seoul, Korea, 13–17 June 2016. No. V07BT31A031.
5. Gao, Q.; Chen, W.; Lu, L.; Huo, D.; Cheng, K. Aerostatic bearings design and analysis with the application to precision engineering: State-of-the-art and future perspectives. *Tribol. Int.* **2019**, *135*, 1–17. [CrossRef]
6. Cappa, S.; Reynaerts, D.; Al-Bender, F. Reducing the Radial Error Motion of an aerostatic Journal Bearing to a Nanometer Level: Theoretical Modelling. *Tribol. Let.* **2014**, *53*, 27–41. [CrossRef]
7. Lie, K.N.; Su, J.C. A numerical analysis of optimum air journal bearings. *J. Aeronaut. Astronaut. Aviation. Ser. A* **2014**, *46*, 49–60.
8. Hosokawa, T.; Somaya, K.; Miyatake, M.; Yoshimoto, S. Static characteristics of aerostatic thrust bearings with multiple porous inlet ports. *J. Tribol.* **2015**, *137*, 1–8. [CrossRef]
9. Zhong, W.; Li, X.; Tao, G.L.; Kagawa, T. Measurement and Determination of Friction Characteristic of Air Flow through Porous Media. *Metals* **2015**, *5*, 336–349. [CrossRef]
10. Belforte, G.; Raparelli, T.; Viktorov, V.; Trivella, A. Permeability and Inertial Coefficients of Porous Media for Air Bearing Feeding Systems. *ASME J. Tribol.* **2007**, *129*, 705–711. [CrossRef]
11. Zhong, W.; Li, X.; Liu, F.H.; Tao, G.L.; Lu, B.; Kagawa, T. Measurement and Correlation of Pressure Drop Characteristics for Air Flow Through Sintered Metal Porous Media. *Transp. Porous Media* **2014**, *101*, 53–67. [CrossRef]
12. Zeng, C.; Wang, W.; Cheng, X.; Zhao, R.; Cui, H. Three-dimensional flow state analysis of microstructures of porous graphite restrictor in aerostatic bearings. *Tribol. Int.* **2021**, *159*, 106955. [CrossRef]
13. Huang, T.Y.; Shen, S.C.; Lin, S.C.; Hsu, S.Y. Pressure Distribution in the Air Film and the Porous Conveyor Air Bearing. *Appl. Mech. Mater.* **2014**, *479*, 380–384. [CrossRef]
14. Cui, H.; Wang, Y.; Yue, X.; Huang, M.; Wang, W. Effects of manufacturing errors on the static characteristics of aerostatic journal bearings with porous restrictor. *Tribol. Int.* **2017**, *115*, 246–260. [CrossRef]
15. Wang, W.; Cheng, X.; Zhang, M.; Gong, W.; Cui, H. Effect of the deformation of porous materials on the performance of aerostatic bearings by fluid-solid interaction method. *Tribol. Int.* **2020**, *150*, 106391. [CrossRef]
16. Van Ostayen, R.A.J.; Van Beek, A.; Munnig-Schmidt, R. Design and Optimization of an Active Aerostatic Thrust Bearing. In Proceedings of the ASPE, Dallas, TX, USA, 14–19 October 2007.
17. Hwang, P.; Khan, P.V. Non-Iterative Finite Element Scheme for Combined Annular-Thrust Porous Aerostatic Bearings Analysis. In Proceedings of the ASME/STLE International Joint Tribology Conference, Denver, CO, USA, 7–10 October 2017.
18. Plante, J.S.; Vogan, J.; El-Aguizy, T.; Slocum, A.H. A design model for circular porous air bearings using the 1D generalized flow method. *Prec. Eng. ISPEN J.* **2005**, *29*, 336–346. [CrossRef]

19. Otsu, Y.; Miyatake, M.; Yoshimoto, S. Dynamic Characteristics of Aerostatic Porous Journal Bearings with a Surface Restricted Layer. *ASME J. Tribol.* **2011**, *133*, 011701-1. [CrossRef]
20. Zhang, X.; Lin, B. Theoretical research on deformation of porous material in air bearing. *Appl. Mech. Mater.* **2012**, *215–216*, 779–784. [CrossRef]
21. Yoshimoto, S.; Kohno, K. Static and Dynamic Characteristics of Aerostatic Circular Porous Thrust Bearings (Effect of the Shape of the Air Supply Area). *ASME J. Tribol.* **2001**, *123*, 501–508. [CrossRef]
22. Su, J.C.T.; Lie, K.N. Rotor dynamic instability analysis on hybrid air journal bearings. *Tribol. Int.* **2006**, *39*, 238–248. [CrossRef]
23. Cui, H.; Wang, Y.; Yue, X.; Li, Y.; Jiang, Z. Numerical analysis of the dynamic performance of aerostatic thrust bearings with different restrictors. Proc. *Inst. Mech. Eng. Part J. Eng. Tribol.* **2019**, *233*, 406–423. [CrossRef]
24. Zang, Y.; Hatch, M.R. Analysis of Coupled Journal and Thrust Hydrodynamic Bearing using Finite Volume Method. *ASME Adv. Inf. Storage Process. Syst.* **1995**, *1*, 71–79.
25. Rahman, M.; Leuthold, H. Computer simulation of a coupled journal and thrust hydrodynamic bearing using a Finite-Element Method. *ASME Adv. Inf. Storage Process. Syst.* **1996**, *2*, 103.
26. Wang, Y.; Wang, Q.J.; Lin, C. Mixed Lubrication of Coupled Journal-Thrust-Bearing Systems Including Mass Conserving Cavitation. *ASME J. Tribol.* **2003**, *125*, 747–755. [CrossRef]
27. Hwang, P.; Khan, P.V. Static Study of the Arc Effect on Annular-Thrust Aerostatic Porous Bearings. In Proceedings of the 5th World Tribology Congress (WTC 2013), Torino, Italy, 8–13 September 2013.

 *applied sciences*

Article

Experimental and Numerical Dynamic Identification of an Aerostatic Electro-Spindle

Federico Colombo *, Luigi Lentini, Andrea Trivella, Terenziano Raparelli and Vladimir Viktorov

Politecnico di Torino, Department of Mechanical and Aerospace Engineering, Corso Duca degli Abruzzi 24, 10129 Turin, Italy; luigi.lentini@polito.it (L.L.); andrea.trivella@polito.it (A.T.); terenziano.raparelli@polito.it (T.R.); vladimir.viktorov@formerfaculty.polito.it (V.V.)

* Correspondence: federico.colombo@polito.it

Featured Application: Aerostatic spindles are used in high-speed micromachining applications. The main goal of the work is to validate the developed non-linear numerical model through the proposed identification procedure and the performed experimental tests.

Abstract: This paper proposes a method to experimentally identify the main modal parameters, i.e., natural frequencies and damping ratios, of an aerostatic spindle for printed board circuit drilling. A variety of methods is applied to the impulse-response function of the spindle in the presence of zero rotational speed and different supply pressures. Moreover, the paper describes the non-linear numerical model of the spindle, which consists of a four-degree-of-freedom (DOF) rigid and unsymmetrical rotor supported by two aerostatic bearings. The main goal of the work is to validate the developed non-linear numerical model through the proposed identification procedure and the performed experimental tests. The comparison proves satisfactory, and the possible sources of uncertainty are conjectured.

Keywords: aerostatic journal bearings; PCB spindles; dynamic identification; logarithmic decrement method; complex exponential method; half-power method; damping factor identification; rigid rotor; modal analysis

Citation: Colombo, F.; Lentini, L.; Trivella, A.; Raparelli, T.; Viktorov, V. Experimental and Numerical Dynamic Identification of an Aerostatic Electro-Spindle. *Appl. Sci.* **2021**, *11*, 11462. https://doi.org/10.3390/app112311462

Academic Editor: Homer Rahnejat

Received: 8 October 2021
Accepted: 29 November 2021
Published: 3 December 2021

Publisher's Note: MDPI stays neutral with regard to jurisdictional claims in published maps and institutional affiliations.

1. Introduction

Gas bearings, owing to their properties, play an essential role in turbochargers, gyroscopes, turbo blowers, gas turbines and micromachining spindles. In particular, micromachining spindles are widely used in surface finishing, printed circuit board (PCB) drilling, micro-processing, micro-milling, micro-grinding and, in general, for micromachining materials with low shear resistance. This is because compared to oil and rolling bearings, gas bearings can reach higher rotation speeds while simultaneously reducing the amount of machine maintenance required.

The aim of research on spindles supported by gas bearings is to develop systems with excellent dynamic stability and stiffness, along with very low runout of the tool, even at very high rotation speeds. Poor damping is one of the main disadvantages of journal gas bearings, which can compromise the accuracy of high-precision machining. Moreover, friction power losses in motors and bearings can be a non-negligible source of heat at high speeds. The contribution of thermally induced deformations of a machine tool can exceed 50% of the total machining error [1], which is very significant in precision milling. To reduce these problems, an efficient and precise refrigeration system is required. Recent developments in manufacturing technologies have made it possible to realize innovative high-precision devices that can even be miniaturized. Research has therefore been enriched by numerous experimental contributions, some of which are discussed below. Machine tools of 4 mm and 1/8" diameters used for micro-milling can now be supported on aerostatic bearings up to 400 krpm and 450 krpm [2,3]. Experiments on dynamic characterization of air-bearing spindles are described in [4,5], where the frequency-response function (FRF) of the spindle is measured. Impact tests have been carried out with an impact hammer [4] or with a custom-designed impact excitation system able to provide excitations up to 20 kHz [5]. Ref. [6] describes a test rig used to measure the dynamic stiffness and damping coefficients of a hemisphere spiral groove hybrid gas bearing. In this rig, the excitation is provided by means of two electromagnetic exciters.

Mathematical models have long been used as a valuable tool to investigate the dynamic performance and stability of high-speed rotors. Early attempts at modelling this kind of system were aimed at gaining a better insight into the half-whirl instability, known today as asynchronous whirl instability [7]. At the time, this was one of the most troublesome problems in journal bearings. The first attempts to actually solve the equations of motion and the gas-lubricated equation simultaneously were reported by Sternlicht [8], Poritsky and Arwas [9]. In their solution, they linearized the equations of motion of a 2 DOF rotor, along with the Reynolds equation, to obtain lubricant pressure distribution. However, in the Reynolds equation, they omitted the time derivative of pressure, and therefore, the resulting quasi-static solutions do not adequately represent the squeeze-film effects. In the same period, the literature presented many other models based on the "*p-h*" linearization of the time-dependent Reynolds equation and the equations of motion of a 4 DOF rigid rotor [10,11]. However, in most cases, these kinds of simplified models cannot replace more costly trial-and-error experimentation since they do not represent a reliable theory.

The results obtained from these models have become ever more accurate with the use of finite-difference and finite-element models, where the perturbation method can be used to compute stiffness and damping coefficients for rotordynamic models. Ref. [12] investigates the impulse response of an ultra-precision raster milling spindle under the simplified hypothesis of constant coefficients. In [13], critical speeds and modal shapes of a micro-spindle machine tools were evaluated, but this analysis is limited because it only considers bearings with constant stiffnesses and neglects their damping properties. A rotordynamic model of an air spindle with a diameter of 1/8″ was considered [14], which involves coefficients that depend on the rotational speed. Here, the critical rotational speeds were calculated as a result of the rotordynamic analysis. A test rig to measure the rotordynamic response of a spindle supported on porous gas bearings is described in [15]. In [16–19], the dependence of stiffness and damping properties on both rotational speed and perturbation frequency is expressed with analytical formulations. With this technique, it is possible to calculate the critical speeds and verify the stability of rigid or flexible rotors. During the design process of a spindle supported on gas bearings, it is of great importance to estimate the conical and cylindrical natural frequencies. A clear design guideline is illustrated in [20] to avoid critical speeds within the operating speed range. The influence of temperature on the dynamic behavior of the spindle is also discussed. Conical and cylindrical modes are investigated in [21] for a non-uniform slot-clearance journal bearing, together with the onset speed of whirl instability.

The orbit method is an alternative to the linearized coefficient analysis employed in the aforementioned papers. It includes the complete nonlinear equations of the rotor and the bearings, which are integrated numerically to obtain the shaft center orbits corresponding to any set of geometrical, running, and initial conditions. This technique essentially uses the computer as an accurate experimental rig; it operates exactly in accordance with the assumed governing equations [22]. This method is currently used to calculate the unbalance response, the onset speed and the non-linear rotor behavior in case of large displacements [23]. Ref. [24] makes use of this method for a spindle with herringbone-grooved journal bearings. Ref. [25] highlights the differences between linear and non-linear analyses of a rotor supported by plain journal bearing. Past research by the authors concerns an electro-spindle of 50 mm shaft dia. supported on aerostatic bearings [26], a textile spindle [27] and a mesoscopic spindle of 10 mm diameter [28]. However, the experimental campaigns to validate the models are not always time- and cost-effective.

This paper proposes a cheap and simple experimental identification procedure for aerostatic spindles. Impulse and static tests are performed to identify the natural frequencies, damping ratio and static stiffness evaluated at the nose of an electro-spindle for PCB drilling. The natural frequencies and damping factors are measured by considering the impulse-response function of the system, whereas the stiffness at the nose spindle is detected through the application of static forces. These spindle features are evaluated at zero rotational speed and with different supply pressures. Moreover, the paper describes the non-linear numerical model of the spindle, which consists of a 4 DOF rigid and unsymmetrical rotor supported by two aerostatic bearings. The equations of motion of the rigid rotor are solved simultaneously with the isothermal time-dependent Reynolds equation through Euler's explicit method to obtain the lubricant pressure distribution. The main goal of the work is to validate the developed non-linear numerical model through the proposed identification procedure and the performed experimental tests.

2. Experimental Setup

2.1. The Prototype

Figure 1a,b show the investigated electro-spindle. During tests, the system was kept in horizontal position by means of a nylon block and a clamp. The spindle is driven by a permanent magnet (PM) synchronous motor that is placed between the rear and front aerostatic journal bearings. A sketch of the rotor-bearings system is shown in Figure 2. The bearings have slightly different diameters: 25 mm (front bearing) and 22 mm (rear bearing). Their axial length is 30 mm. They are supplied by means of two series of supply orifices of diameter d_s = 0.15 mm, located at 7 mm from the borders of the bearings. Each series includes 12 holes equally spaced along the circumferential direction. Downstream each supply hole, a small pocket of dia. 1.1 mm and depth 0.2 mm is present. Considering the manufacturing tolerances, the radial air-film thicknesses are h_f = 17.5 $\pm$ 1.5 μm and h_r = 21.5 $\pm$ 1.5 μm on the front and rear bearings, respectively. Table 1 lists the mass properties of the rotor. The spindle is equipped with a water-cooling system to ensure temperature stability during operation.

Table 1. Mass properties of rotor.

Rotor mass:	m_r = 382 g
Transversal inertia moment:	I = 666.1 $\times$ 10^{-6} kg m^2
Polar inertia moment:	I_p = 31.1 $\times$ 10^{-6} kg m^2

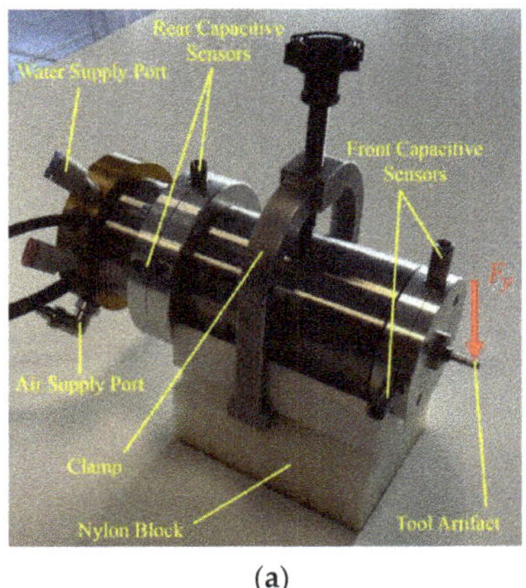

(**a**)

(**b**)

Figure 1. (**a**) Photo of the electro-spindle in horizontal attitude; (**b**) photo of the test setup.

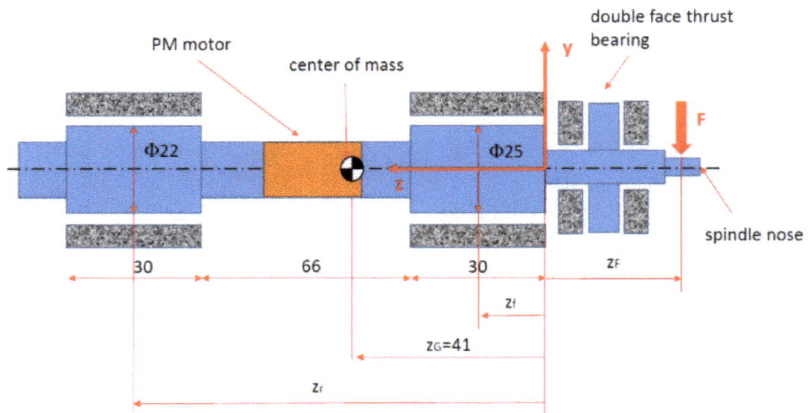

Figure 2. Sketch of the rotor-bearings system.

2.2. Shaft-Displacement Detection

Figure 2 shows the cartesian reference system, $Oxyz$, used to define the rotor center of mass and its position with respect to the bearings. The origin, O, is located at the front edge of the front journal bearing. Axis z is along the axial direction and directed from the front to the rear bearing. The axial coordinates of journal centers are indicated with z_f and z_r for front and rear bearings, respectively.

The rotor displacement is detected by means of two couples of capacitance probes located at $z_1 = -21$ mm (front plane) and $z_2 = 91.5$ mm (rear plane). The center of mass of the shaft is located at $z_G = 41$ mm. The shaft center-of-mass displacements and the rotations about x and y axes are obtained from the sensor readings on front (x_1, y_1) and rear planes (x_2, y_2), as below:

$$
\begin{aligned}
x_G &= \frac{x_1(z_2 - z_G) - x_2(z_1 - z_G)}{z_2 - z_1} \\
y_G &= \frac{y_1(z_2 - z_G) - y_2(z_1 - z_G)}{z_2 - z_1} \\
\vartheta_y &= \frac{x_2 - x_1}{z_2 - z_1} \\
\vartheta_x &= \frac{y_1 - y_2}{z_2 - z_1}
\end{aligned}
\tag{1}
$$

The shaft displacement at the spindle nose ($z = z_{nose} = -40$ mm) can be extrapolated using Equation (2):

$$
\begin{aligned}
x_{nose} &= \frac{x_1(z_2 - z_{nose}) - x_2(z_1 - z_{nose})}{z_2 - z_1} \\
y_{nose} &= \frac{y_1(z_2 - z_{nose}) - y_2(z_1 - z_{nose})}{z_2 - z_1}
\end{aligned}
\tag{2}
$$

2.3. Static Stiffness on the Nose

The static stiffness on the nose was evaluated by extrapolating the displacement produced through the application of a 30 ± 1 N load at $z = z_{nose}$. A dynamometer was employed to impose the load, and the spindle displacement at $z = z_{nose}$ was computed by using Equation (2). Table 2 lists the resulting stiffness, measured at different supply pressures.

Table 2. Resulting static stiffness on the nose.

p_s (MPa)	k_{nose} (N/μm)
0.3	1.14
0.5	2.9
0.7	4.0
0.9	4.6

2.4. Experimental Characterization at Null Rotational Speed

The damped natural frequencies, ω_d, and the damping factors, ζ, were experimentally investigated by means of impulse response tests. A vertical impulse was applied to the nose in correspondence of the tool artifact, and the spindle trajectory was detected by the capacitive sensors (capaNCDT CS05 by Microepsilon) (see Figure 1a). The tests were performed at null rotational speed and absolute supply pressures p_s = 0.3, 0.5, 0.7 and 0.9 MPa. Each test was repeated four times, and signals were sampled with sampling frequency f_s = 50 kHz.

The spectra of signals y_1 and y_2, measured by sensors on the front and rear measuring planes, were evaluated through fast Fourier transform (FFT) in order to visualize the number of modes captured and their frequencies. Figure 3 shows one of the computed spectra (at 0.5 MPa supply pressure), while the related time signals are depicted in Figure 4. In all the investigated cases, only one rigid mode shape is visible. Meanwhile, the higher frequency can be attributed to the flexural mode of the spindle since it does not depend on the supply pressure of the air bearings. As can be seen from Figure 4, the first mode shape is conical since y_1 and y_2 present a $180°$ phase shift.

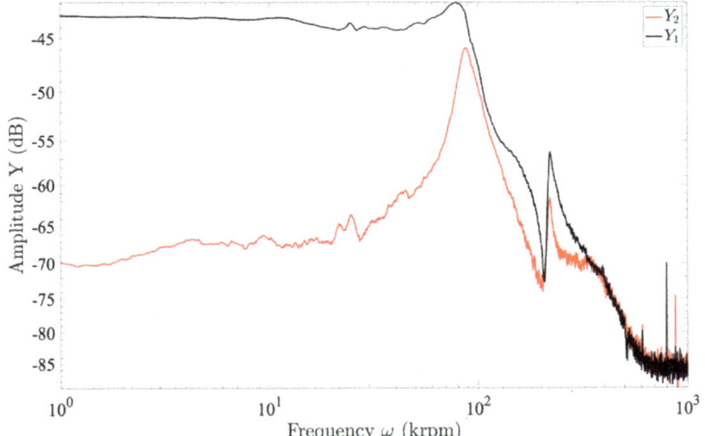

Figure 3. Example of spectra measured at 0.5 MPa supply pressure.

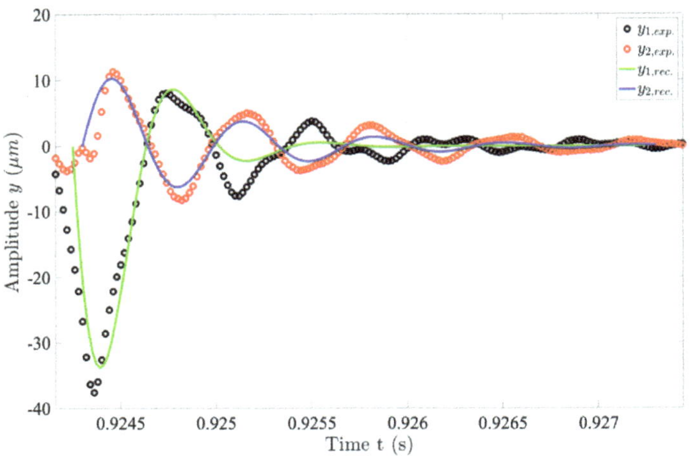

Figure 4. Example time signals at 0.5 MPa supply pressure.

2.4.1. Single-DOF Modal-Identification Analysis through LDM and HPM

In view of these considerations, as a first approach, two single-DOF methods were employed to identify the modal parameters of the system: the logarithmic decrement method (LDM) [29] and the half-power method (HPM) [30]. Table 3 lists the mean values of the estimations of the damped frequencies, ω_d, and the damping ratios, ζ, computed through HPM and LDM. The values of the damped natural frequencies increase with the supply pressure, whereas the damping ratios exhibit an opposite trend. The accuracy of these estimations was verified by comparing the experimental signals with those reconstructed through the identified modal parameters (see Figure 4) by making use of the following 1 DOF analytical formula:

$$y_{rec}(t) = [a\cos(\omega_d t) + b\sin(\omega_d t)] \cdot e^{-\zeta\omega_n t}$$
$$a = y_{rec}(0)$$
$$b = \frac{\dot{y}_{rec}(0) + \zeta\omega_n y_{rec}(0)}{\omega_d} \tag{3}$$

where constants a and b depend on the initial conditions, and the undamped natural frequency is $\omega_n = \omega_d/\sqrt{1-\zeta^2}$.

Table 3. Experimental natural frequency and damping ratio of signal y_2 at different supply pressures; the results were obtained with HPM and LDM.

p_s (MPa)	HPM		LDM	
	ω_d (krpm)	ζ	ω_d (krpm)	ζ
0.3	70.23	0.2611	72.82	0.3928
0.5	87.63	0.1464	86.50	0.2347
0.7	95.42	0.1084	92.60	0.1497
0.9	100.3	0.1125	97.64	0.2325

Unfortunately, as can be seen from Figure 4, the reconstructed signals do not exhibit a satisfactory degree of accuracy if the estimated parameters are not manually corrected with a trial-and-error procedure. This mismatch was mainly due to the fact that in some cases, the analyzed modes presented a relatively high damping factor (small number of oscillations in the time signal) and a flat peak in the spectrum. In fact, despite the large amount of energy provided by hammering the spindle nose, in most cases, the rotor stopped after a few oscillations. To overcome these problems, the estimations were repeated by means of the least-squares complex exponential method (LSCEM).

2.4.2. Multi-DOF Modal Analysis with LSCEM

This is a multi-DOF method that works in the time domain [31]. It is based on the impulse-response function (IRF) of the MDOF system with N couples of complex and conjugate poles:

$$h_{ij}(t) = \sum_{r=1}^{2N} A_{r,ij} \cdot e^{s_r t} \tag{4}$$

where $s_r = \zeta_r\omega_{n,r} \pm i\,\omega_{n,r}\sqrt{1-\zeta_r^2}$ and $A_{r,ij}$ are the rth poles and the modal constants of the system and i and j refer to the nodes where the impulse is applied and where the system response is measured, respectively. The advantage of this method is that the nonlinear solution for the eigenvalues is computationally very straightforward since it performs an exponential fitting that requires no starting values. To obtain a solution, it is only necessary to supply the algorithm with the impulse-response function, $h_{ij}(t)$, and the order of the model (N). A particular solution strategy was adopted in these cases. In order to provide the algorithm with the possibility to fit-noisy signals, the order of the model was chosen equal to 20, and frequency stabilization diagrams were used to distinguish the "computational modes" from the real ones. This kind of diagram was obtained by iterating the LSCEM from a minimum (3) to a maximum order (20) and superimposing the obtained frequencies on the obtained spectrum.

In order to obtain a cleaner stabilization diagram, before plotting the obtained damped frequencies, ω_d, they were filtered by considering only the first 4 modes presenting the higher modal constant ($\omega_{1,N}$, $\omega_{2,N}$, $\omega_{3,N}$ and $\omega_{4,N}$). Figures 5 and 6 show the spectrum and time signal reconstructed by using the modal parameters obtained from this procedure. In Figure 5, the results correspond to the "+" symbols, and they are coloured in blue, red, cyan and black from the higher to the lower modal constant. A further filter was imposed by considering a threshold of 5% on the values of the normalized modal constants in an attempt to exclude the computational modes. The results that satisfied this threshold are circled in green (see Figure 5). Figure 5 also shows the comparison between the experimental spectrum and that reconstructed through the LSCEM. Figure 6 shows the same comparison in the time domain.

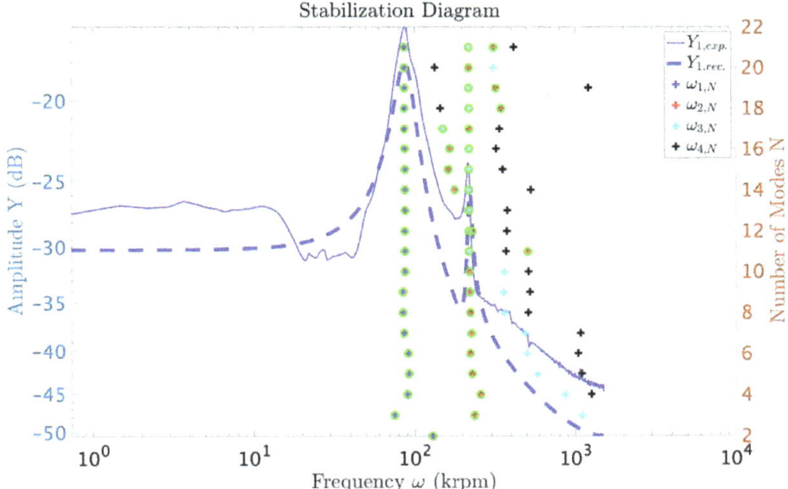

Figure 5. The experimental and reconstructed spectra at 0.5 MPa supply pressure.

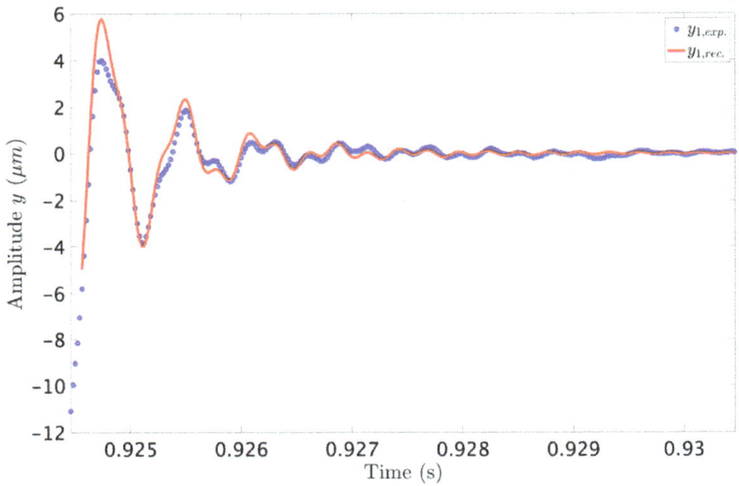

Figure 6. The experimental and reconstructed time signals at 0.5 MPa supply pressure.

As can be seen, in this case, the accuracy of the method is satisfactory. The results obtained with the LSCEM method are summed up in Table 4, where the damped natural frequencies and damping ratios are reported, together with their phase, ϕ, and modal constant, A. Only the first rigid mode was considered, as the second mode exhibited a natural frequency of about 220 krpm, which was almost constant with the supply pressure and had modal constants much smaller than the first mode. The evaluation of the phase shifts, $\Delta\phi = |\phi(y_1) - \phi(y_2)|$, confirms that the mode identified is of the tilting type.

Table 4. Experimental natural frequency and damping ratio of signals y_1 and y_2 at different supply pressures (only the first rigid mode is reported); the results were obtained with LSCEM.

p_s (MPa)	Signal	ω_d (krpm)	ζ (−)	ϕ (deg)	A_1 (−)
0.3	y_2	73.44	0.1483	3.474	3.400
	y_1	70.21	0.2578	169.8	7.963
0.5	y_2	89.50	0.1072	9.420	5.305
	y_1	87.23	0.1632	178.0	7.673
0.7	y_2	99.75	0.00762	−3.336	5.223
	y_1	95.46	0.1060	153.6	8.461
0.9	y_2	103.8	0.07330	19.69	3.150
	y_1	107.0	0.09230	−154.6	4.102

2.4.3. LSCEM vs. LDM and HPM (Experimental)

The damped frequencies obtained with the 1 DOF methods are in good accordance with those obtained with LSCEM; the error is less than 4%. The damping factors, apart from cases with $p_s = 0.3$ MPa, differ by less than 36% for HPM and 61% for LDM.

3. Numerical Model

In this section, the numerical model of the spindle described in Section 2.1 is presented. The effect of the double-face thrust bearing is not taken into account, as it is negligible in the determination of radial and tilting stiffness with respect to the contribution of journal bearings.

The time-dependent Reynolds equation is considered in the fluid domain of journal bearings:

$$\frac{\partial}{\partial z}\left(\frac{ph^3}{12\mu RT}\frac{\partial p}{\partial z}\right) + \frac{\partial}{r\partial\vartheta}\left(\frac{ph^3}{12\mu RT}\frac{\partial p}{r\partial\vartheta}\right) + g_{in} = \frac{\omega r}{2RT}\frac{\partial(ph)}{r\partial\vartheta} + \frac{1}{RT}\frac{d(ph)}{dt} \qquad (5)$$

where g_{in} is the input flow per unit surface that crosses the supply orifices. The input flow is estimated using the analytical expression of the supply hole's discharge coefficient [32] without considering the Reynolds number dependence:

$$c_d = 0.85\left(1 - e^{-8.2\frac{h}{d_s}}\right) \qquad (6)$$

where d_s is the supply hole's diameter and h is the local air gap under the orifice. The RE is discretized with finite difference technique, and the pressure distribution at time $t + 1$ is obtained from the previous pressure distribution with the Euler explicit method:

$$p_{ij}^{t+1} = p_{ij}^t + \Delta t \cdot f\left(p^t, p^{t+1}\right) \qquad (7)$$

where f is a non-linear function that involves the pressure values in node i,j and in the adjacent nodes. The pocket downstream of each supply orifice is also considered, as it influences the dynamic characteristics of bearings. Once pressure p_{ij} is calculated at time $t + 1$, this value is immediately used to calculate $p_{i+1,j}$ at the same time. This improves the stability of the numerical scheme, as it facilitates the transmission of the information in the adjacent nodes. The film thickness depends on the rotor center-of-mass position (x_G, y_G) and on the rotor axis orientation $(\vartheta_x, \vartheta_y)$:

$$h(z, \vartheta) = h_0 - \left[x_G + \vartheta_y(z - z_G)\right] \cos \vartheta - \left[y_G - \vartheta_x(z - z_G)\right] \sin \vartheta \tag{8}$$

Once the pressure distribution is calculated, the shear stress on the rotor surface is given by:

$$\tau = \frac{h}{2} \frac{\partial p}{r \partial \vartheta} + \mu \frac{\omega r}{h} \tag{9}$$

The 4 DOF equations of motion for a rigid rotor are considered:

$$\begin{aligned}
m_r \ddot{x}_G - F_{cx} - F_{ext,\, x} &= m_r e \omega^2 \cos \omega t \\
m_r \ddot{y}_G - F_{cy} - F_{ext,\, y} &= m_r e \omega^2 \sin \omega t \\
I \ddot{\vartheta}_x + I_p \omega \dot{\vartheta}_y - M_{cx} - F_{ext,\, y}(z_G - z_{nose}) &= (I_p - I)\gamma \omega^2 \sin(\omega t + \varphi_1) \\
I \ddot{\vartheta}_y - I_p \omega \dot{\vartheta}_x - M_{cy} + F_{ext,\, x}(z_G - z_{nose}) &= (I - I_p)\omega^2 \gamma \cos(\omega t + \varphi_1)
\end{aligned} \tag{10}$$

where e and γ are the static and the dynamic rotor unbalances, respectivelye; φ_1 is the angle that locates the dynamic unbalance plane with respect to the static unbalance plane; $F_{ext,x}$ and $F_{ext,y}$ are the external-force components acting on the rotor at coordinates z_{nose}; F_{cx}, F_{cy}, M_{cx} and M_{cy} are the reaction forces and moments of the bearings, calculated by integrating the pressure distribution and the shear stress in the fluid domain.

The moments are expressed with respect to the rotor center of mass:

$$\begin{bmatrix} F_{cx} \\ F_{cy} \end{bmatrix} = \int_0^L \int_0^{2\pi} \left(-p \begin{bmatrix} \cos \vartheta \\ \sin \vartheta \end{bmatrix} + \tau \begin{bmatrix} \sin \vartheta \\ -\cos \vartheta \end{bmatrix} \right) r\, d\vartheta\, dz \tag{11}$$

The moments are expressed with respect to the rotor center of mass:

$$\begin{bmatrix} M_{cx} \\ M_{cy} \end{bmatrix} = \int_0^L \int_0^{2\pi} \begin{bmatrix} (z - z_G)(p \sin \vartheta + \tau \cos \vartheta) \\ (z - z_G)(-p \cos \vartheta + \tau \sin \vartheta) \end{bmatrix} r\, d\vartheta\, dz \tag{12}$$

The time integration of these equations of motion are carried out with the first order Euler method:

$$q_{t+1} = q_t + \dot{q}_t \Delta t \tag{13}$$

where q is the state-space vector defined by

$$q = \begin{bmatrix} x_G & y_G & \vartheta_x & \vartheta_y & \dot{x}_G & \dot{y}_G & \dot{\vartheta}_x & \dot{\vartheta}_y \end{bmatrix}$$

The algorithm performed for the coupled integration of the ODE and the PDE system is described below. After assuming an initial pressure distribution, for each time iteration, the following steps are followed:

1. The reaction forces and moments are calculated using Equations (11) and (12);
2. The center-of-mass accelerations and the angular accelerations of the rotor are computed from the equations of motion (10);
3. The state-space vector at time $t + 1$ is obtained from Equation (13);
4. The film thickness at time $t + 1$ is updated from Equation (8);
5. The pressure and shear-stress distributions at time $t + 1$ are updated by solving RE (2) and using Equation (9);
6. Go back to point 1.

This method is known as the "orbit method" and has the advantage of also taking into account the non-linearities of bearings [22]. Conversely, it is more time-consuming compared to linearized methods.

For static problems, the rotor position is fixed, and only the pressure distribution is considered time-dependent. In this case, only steps 1 and 5 are considered in order to find the steady-state solution. The iterative process is stopped when the load-capacity relative variation decreases below a given tolerance:

$$\left| \frac{F_c^{t+1} - F_c^t}{F_c^t} \right| < 10^{-6} \tag{14}$$

4. Numerical Results

The numerical model developed was validated using both literature benchmarks and numerical results. In particular, the numerical load capacity, $\overline{W}$, and attitude angle, Φ, were compared with the analytical values for plain dynamic journal bearings. In the comparison, the static stiffness of the spindle measured in correspondence of the nose was also considered, together with the damped natural frequencies and the damping ratio at null rotational speed.

4.1. Static Validation with Analytical Solution of Plain Dynamic Journal Bearings

The mathematical model was validated in static conditions, comparing it with the analytical solution that exists for the plain dynamic journal bearing [7]. The dimensionless load capacity is expressed with a real and imaginary parts, which represent the in-phase and the out-of-phase components with respect to the line of centers:

$$\overline{W} = \frac{W}{\epsilon p_a L D} = \frac{i\Lambda}{1 + i\Lambda} \frac{\pi}{2} \left[1 - \frac{\tanh\left(\sqrt{1 + i\Lambda \frac{L}{D}}\right)}{\sqrt{1 + i\Lambda \frac{L}{D}}} \right] \tag{15}$$

where

$$\Lambda = \frac{6\mu\omega}{p_a} \left(\frac{r}{C}\right)^2 \tag{16}$$

is the bearing number, and

$$\epsilon = \frac{e}{C} \tag{17}$$

the eccentricity ratio. The front and rear journal bearings were simulated with no supply orifices for comparison with the dynamic-bearing analytical solution. Figure 7 shows a perfect match, both regarding the load capacity and the attitude angle, Φ.

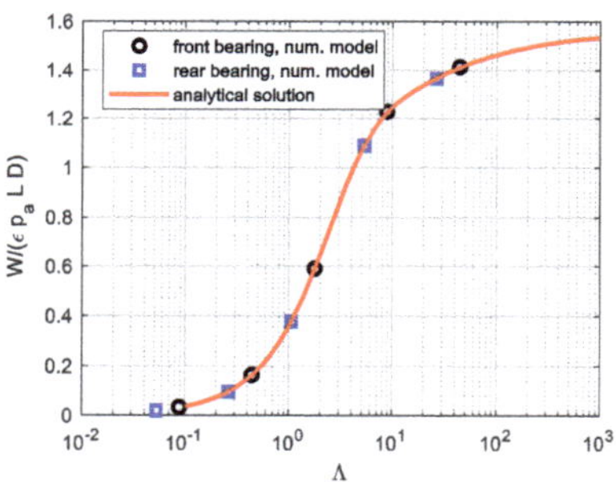

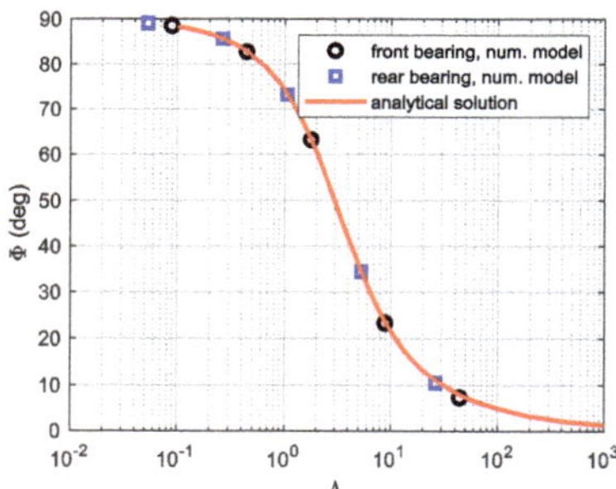

Figure 7. Dimensionless load capacity and attitude angle of plane dynamic journal bearings; comparison between the analytical formulation and numerical results.

4.2. Choice of Grid Resolution

The influence of the grid-point number on the numerical solution is evaluated both in static and dynamic conditions. A total of 31 nodes were chosen along the axial direction for each bushing. Along the circumferential direction, 48 or 96 nodes were considered. Figures 8–10 compare the pressure distribution, load capacity and air consumption obtained with these computational grids in the presence of an eccentricity of $e = 1$ μm. As can be seen, the difference is minimal. As a result of these considerations, a 31×48 grid was used.

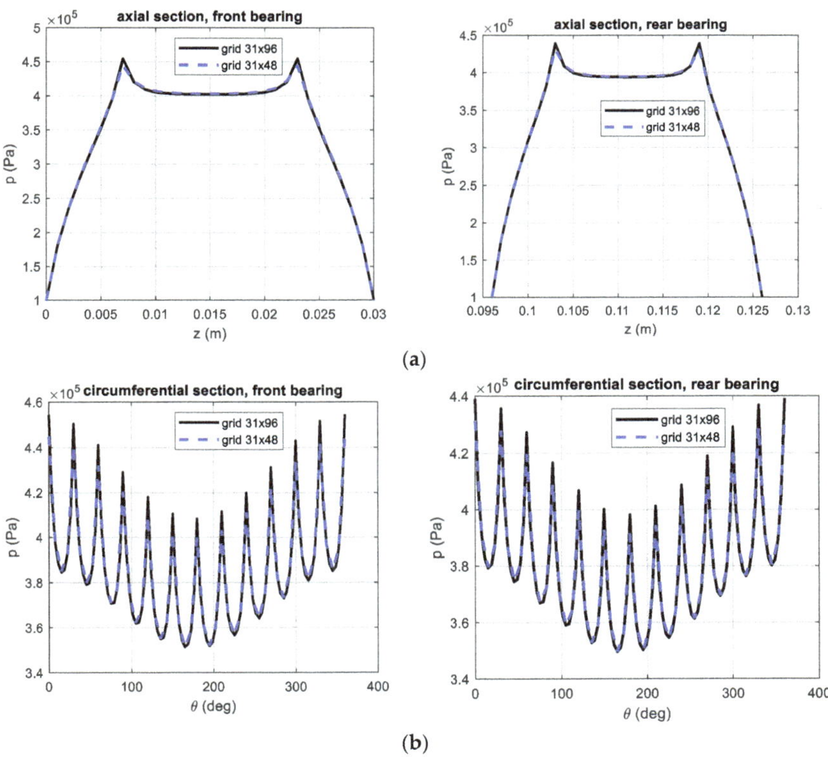

Figure 8. Axial (**a**) and circumferential (**b**) pressure distributions obtained with different numbers of grid points along the circumferential direction (48 and 96); $p_s = 0.5$ MPa.

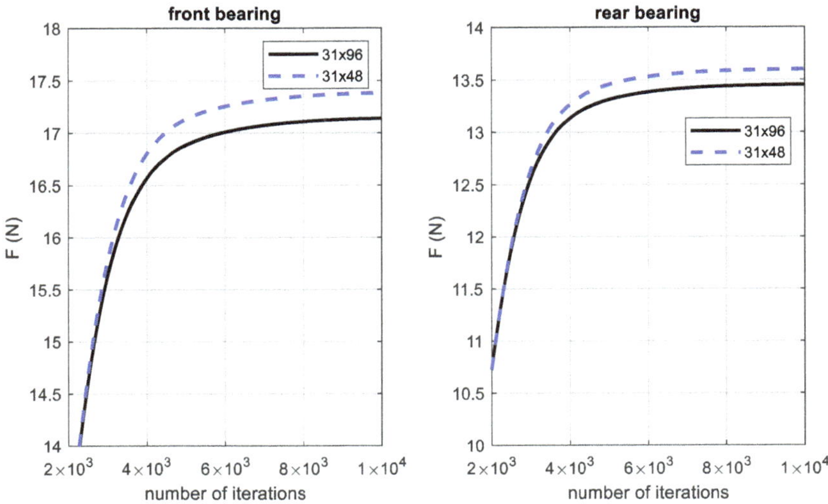

Figure 9. Load capacity vs. number of iterations obtained with different grid numbers along the circumferential direction.

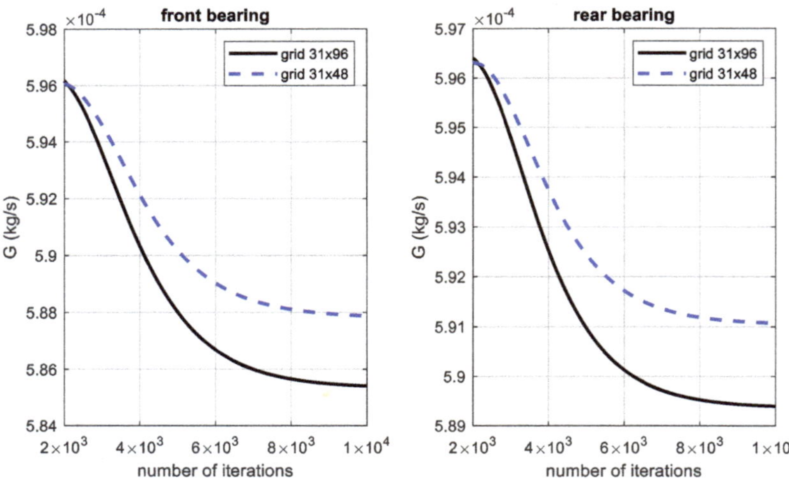

Figure 10. Air flow vs. number of iterations obtained with different grid numbers along the circumferential direction.

4.3. Static Stiffness of Journal Bearings

The stiffness of journal bearings at zero speed was obtained from static simulations. A given displacement was imposed on the shaft (1 μm on radius), and the reaction forces in the bearings were computed in order to estimate the radial stiffness, k_f and k_r, of the front and rear bearings, respectively. The stiffness values are listed in Table 5, considering minimal and maximal radial clearances and taking into account the manufacturing tolerances, as shown in Section 2.1. Of course, stiffness values increase with supply pressure and decrease with increasing air gaps. The front bearing shows an increment of stiffness with the air gap only with p_s = 0.3 MPa. The following explanation can be given after investigating the pressure distribution: the reason is that the front bearing is near the saturation condition; that is to say that the pressure drop through the input orifices is quite small (less than 0.03 MPa).

Table 5. Radial stiffness of journal bearings at different supply pressures and radial clearances.

p_s (MPa)	h_f = 16 μm, h_r = 20 μm		h_f = 19 μm, h_r=23 μm	
	k_f (N/μm)	k_r (N/μm)	k_f (N/μm)	k_r (N/μm)
0.3	4.51	4.57	5.53	4.22
0.5	12.6	9.37	11.8	7.45
0.7	20.3	13.0	16.7	9.5
0.9	27.0	15.8	20.6	10.9

The tilting stiffness of each journal bearing, calculated with respect to its center, is also evaluated in order to investigate whether its contribution to the overall tilting stiffness of the rotor bearings system is negligible or not. Table 6 compares the contribution of each bearing to the tilting stiffness obtained from the radial stiffness (left column) with the tilting stiffness of each bearing with respect to its center (right column). From this comparison, it is evident that in this case, this last bearing can be neglected so the overall tilting stiffness can be estimated from the radial stiffness of each journal bearing.

Table 6. Comparison between different contributions to tilting stiffness; results refer to geometrical configuration with h_f = 19 μm, h_r = 23 μm and supply pressure p_s = 0.7 MPa.

Tilting Stiffness of Journal Bearing with Respect to the Shaft Center of Mass		Tilting Stiffness of Journal Bearing Respect to Its Center	
$k_f \cdot (z_G - z_f)/\theta_x$ (Nm/rad)	$k_r \cdot (z_G - z_r)/\theta_x$ (Nm/rad)	k_{ϑ_f} (Nm/rad)	k_{ϑ_r} (Nm/rad)
33507	57166	925	534

4.4. Stiffness on the Nose

The overall stiffness, k_{nose}, measured on the nose at $z = z_F = -40$ mm, is a function of the bearing stiffness according to:

$$k_{nose} = k_f \frac{z_r - z_f}{z_r - z_F} \qquad (18)$$

Table 7 lists the radial stiffness on the nose extrapolated from the journal bearings stiffness calculated in the centered position, with 1 μm rotor eccentricity. In case an external radial load of 30 N is applied on the nose, the eccentricities on bearings are greater and the resulting nose stiffness is lower. Table 8 lists these values.

Table 7. Radial stiffness on the nose at different supply pressures and radial clearances, calculated at small eccentricity (1 μm).

p_s (MPa)	k_{nose} (N/μm) h_f = 16 μm, h_r=20 μm	k_{nose} (N/μm) h_f = 19 μm, h_r=23 μm
0.3	2.87	3.51
0.5	8.01	7.50
0.7	12.9	10.6
0.9	17.1	13.1

Table 8. Radial stiffness of the nose at different supply pressures and radial clearances; these values were obtained with a 30 N load applied on the nose.

p_s (MPa)	k_{nose} (N/μm) h_f = 16 μm, h_r=20 μm	k_{nose} (N/μm) h_f = 19 μm, h_r=23 μm
0.3	1.32	1.29
0.5	4.02	3.79
0.7	6.59	5.36
0.9	8.82	6.52

4.5. Simplified Natural Frequencies

At null rotational speed, in case the damping effects are neglected, the cylindrical natural frequency is expressed by:

$$\omega_{cyl} = \sqrt{\frac{k_f + k_r}{m_r}} \qquad (19)$$

The conical natural frequency is given by:

$$\omega_{con} = \sqrt{\frac{\left(k_f l_f^2 + k_r l_r^2\right)}{I}} \qquad (20)$$

where l_f and l_r are the distances between the rotor center of mass and the centers of the front and the rear journal bearings, respectively. In this formulation, the eventual tilting stiffness of each journal bearing is neglected. Table 9 reports the undamped natural frequencies of the rigid rotor-bearings system at null rotational speed.

Table 9. Undamped natural frequencies at different supply pressures and radial clearances.

p_s (MPa)	$h_f = 16$ μm, $h_r = 20$ μm		$h_f = 19$ μm, $h_r = 23$ μm	
	ω_{cyl} (krpm)	ω_{con} (krpm)	ω_{cyl} (krpm)	ω_{con} (krpm)
0.3	46.6	58.9	48.2	57.7
0.5	72.4	86.2	67.8	78.0
0.7	89.2	102.8	79.1	88.9
0.9	101.1	114.3	86.7	95.9

4.6. Choice of the Time Step

The optimal time step, dt, must be chosen for simulations in order to guarantee a stable solution and independence of the dynamic solution on the time step. The entity of the damping factor was found to be dependent on the time step, dt, and on the grid refinement. Figure 11 compares the values obtained for the damping factor at different rotational speeds and time steps, dt, with a 31×96 grid. As shown, the convergence is reached with a time step of $dt = 10^{-8}$ s, and a further decrease does not significantly change the damping factor. The effect of the grid on the damping factor is visible in Figure 12. When doubling the grid nodes along the circumferential direction, it is sufficient to halve the time step, dt, in order to obtain a very similar solution. As a result of these considerations, a 31×48 grid with $dt = 2 \times 10^{-8}$ s was chosen for dynamic simulations, as it was found to be a good compromise between accuracy and computational time.

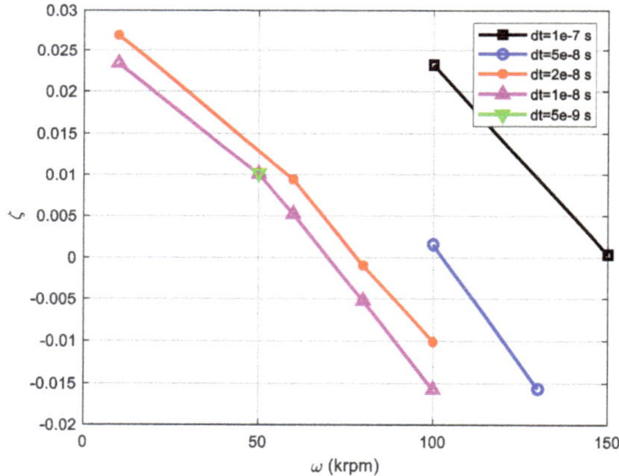

Figure 11. Damping ratio vs. rotational speed obtained with different time steps; the grid is 31×96.

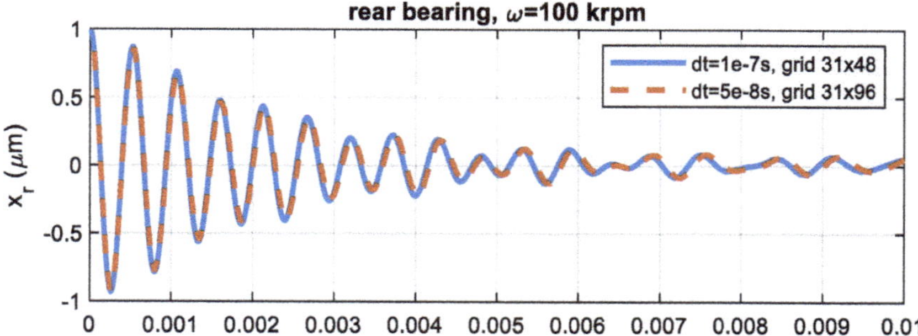

Figure 12. *Cont.*

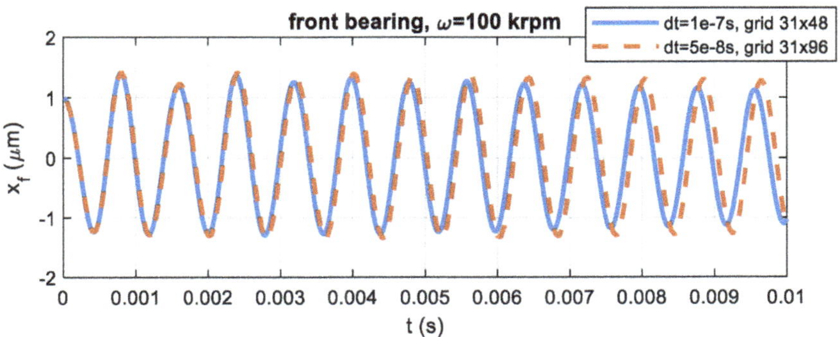

Figure 12. Comparison of vibrations obtained on front and rear bearings with 31×96 and 31×48 grids; the two curves match fairly well in case the time step is halved when doubling the grid points.

4.7. Damped Natural Frequencies and Damping-Factor Identification

The damped natural frequencies and the damping factors were evaluated for the displacements along the x and y directions by the rotor in correspondence of planes at $z = 0$ (at the right edge of the front journal bearing) and at $z = z_L = 126$ mm (at the left edge of the rear journal bearing). Moreover, to better identify both the cylindrical and the conical modes, the center-of-mass displacement, y_G, and the tilting angles, ϑ_x and ϑ_y, were considered. Two different initial conditions were simulated since the frequency content of the spectra of the system response may depend on how the system is excited. Condition 1 is defined by:

$$x(z = 0, t = 0) = 0 , \; y(z = 0, t = 0) = 0$$
$$x(z = z_L, t = 0) = 0, \;\; y(z = z_L, t = 0) = 0$$
$$\dot{x}(z = 0, t = 0) = 0 , \; \dot{y}(z = 0, t = 0) = -0.01 \text{ m/s}$$
$$\dot{x}(z = z_L, t = 0) = 0 , \; \dot{y}(z = z_L, t = 0) = 0$$

Condition 1 was adopted in an attempt to reproduce the performed experimental tests. Vertical impulses were applied to the spindle nose, and the rotor oscillations were measured on the rear and front plane. Since the results obtained with the initial conditions (condition 1) were mainly governed by the conical mode, a second condition (condition 2) was simulated in an attempt to better figure out the cylindrical mode-shape of the rotor.

$$x(z = z_G, t = 0) = 1 \text{ μm}, \;\; y(z = z_G, t = 0) = 0$$
$$\vartheta_x(t = 0) = 0, \;\; \vartheta_y(t = 0) = 0$$
$$\dot{x}(z = z_G, t = 0) = 0 , \; \dot{y}(z = z_G, t = 0) = -0.01 \text{ m/s}$$
$$\dot{\vartheta}_x(t = 0) = 0 , \; \dot{\vartheta}_y(t = 0) = 0$$

Condition 2 was chosen to simultaneously produce a whirl and a translation of the rotor. Due to the non-symmetry of the rotor-bearings system, both modes were visible both with initial condition 1 and with condition 2. The frequency content of rotor displacements puts in evidence the general presence of two natural frequencies, as seen in Figure 13. The prominence of one with respect to the other depends on the initial condition.

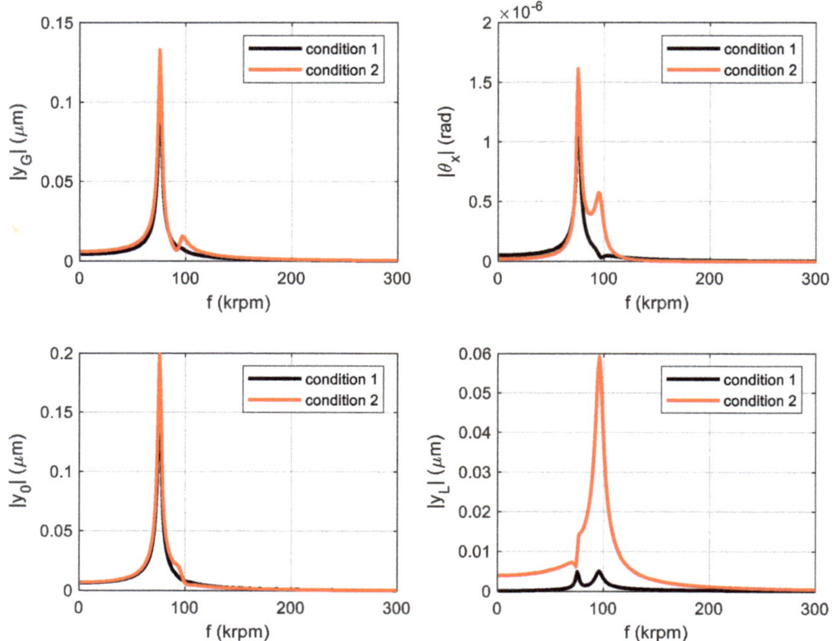

Figure 13. Examples of frequency spectra of signals y_G, y_0, θ_x and y_L obtained from the orbit method.

4.7.1. Logarithmic Decrement Method (LDM)

Although the LSCEM proved to be the most reliable method, for the sake of completeness, the LDM method was also used to identify the main damping factors (with the largest modal constant) and natural frequencies of the numerical signals y_0, y_L, y_G and θ_x. The LDM was used, considering time intervals ΔT of 5 and 10 ms. The resulting values of the damping factors were compared to evaluate the sensitivity of the method with respect to the signal length. The main frequency of each signal was also evaluated by measuring the average period of the oscillations present in the considered time intervals. This frequency can be compared with the peaks visible in the related spectrum. Tables A1–A4 in Appendix A list both the natural frequencies resulting from the FFT analysis and the frequencies and the damping factor of the signal resulting from the logarithmic decrement method. When the oscillation was highly damped ($p_s = 0.3$ and 0.5 MPa), only a 5 ms time window was considered. The following considerations can be made based on the data analysis:

- The frequency obtained from the average period of the oscillation is quite similar to one of the two frequencies detected in the FFT spectrum; in particular, the higher frequency is visible only on the rear plane (signal y_L), while the lower frequency is present on the other signals (y_0, y_G and ϑ_x);
- in most cases, the application of LDM to signals with durations of 5 and 10 ms does not influence the results;
- depending on the initial condition, one of the two modes prevails;
- due to the short duration of the vibration at $p_s = 0.3$ MPa, which is highly damped, the FFT algorithm does not provide a good estimation of the frequency spectra;
- the frequency increases with the supply pressure, while the damping factor decreases;
- by increasing the air gap, the frequency decreases, as well as the damping factor.

4.7.2. LSCE Method

The frequency content of the numerical signals y_L and y_0 was investigated by adopting a procedure similar to that adopted in Section 2.4. To obtain more reliable frequency-stabilization diagrams, the adopted procedure was slightly different than that used for the experimental data. In this case, to favor the exponential fitting, the model order used in the LSCEM was 80, and a random noisy signal was added to the numerical signals [31]. Figures 14 and 15 show the comparison obtained between the original and reconstructed data. The damped frequencies and damping factors evaluated on signals y_0 and y_L are listed in Table 10, together with their phase shift and modal constants.

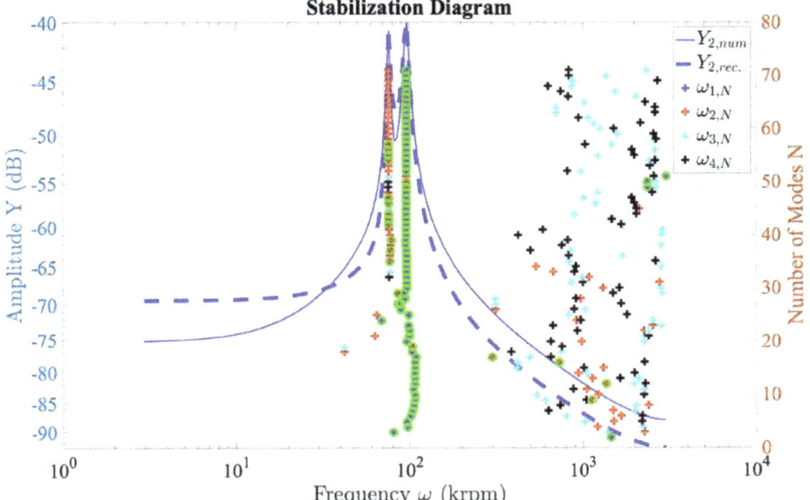

Figure 14. Stabilization diagram.

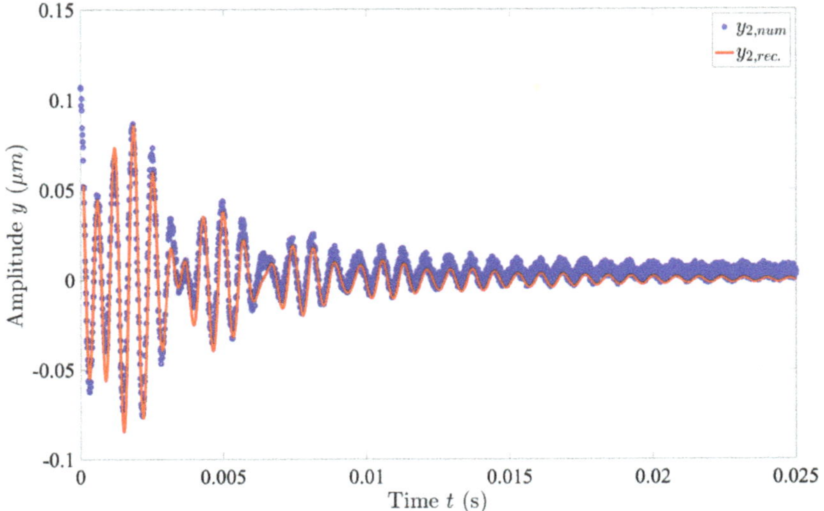

Figure 15. Comparison in time domain of the reconstructed signal with the numerical signal.

Table 10. Modal parameters resulting from LSCEM.

p_s (MPa)	h_f/h_r	Signal	$\omega_{d,1}$ (krpm)	ζ_1 (-)	φ_1 (deg)	A_1 (-)	$\omega_{d,2}$ (krpm)	ζ_2 (-)	φ_2 (deg)	A_2 [-]
0.3	16/20	y_L	80.51	0.1675	−19.07	0.0737	n.a.	n.a.	n.a.	n.a.
		y_0	58.28	0.3427	−54.14	0.9738	n.a.	n.a.	n.a.	n.a.
0.3	19/23	y_L	70.00	0.1556	−18.80	0.1224	n.a.	n.a.	n.a.	n.a.
		y_0	50.00	0.2044	−70.50	1.115	n.a.	n.a.	n.a.	n.a.
0.5	16/20	y_L	100.6	0.0804	−21.40	0.0618	n.a.	n.a.	n.a.	n.a.
		y_0	74.82	0.1394	−76.58	0.7109	n.a.	n.a.	n.a.	n.a.
0.5	19/23	y_L	86.97	0.0620	−21.69	0.0710	66.22	0.0667	100.1	0.0377
		y_0	65.86	0.0629	82.46	0.7587	n.a.	n.a.	n.a.	n.a.
0.7	16/20	y_L	113.3	0.0365	21.8	0.029	87.83	0.0604	−50.2	0.0135
		y_0	87.69	0.0652	115.17	0.3245	n.a.	n.a.	n.a.	n.a.
0.7	19/23	y_L	96.09	0.0356	−23.30	0.0496	75.80	0.0193	79.62	0.0218
		y_0	75.81	0.0166	−85.73	0.6339	n.a.	n.a.	n.a.	n.a.
0.9	16/20	y_L	122.3	0.0124	4.855	0.0251	97.46	0.0281	−14.7	0.0104
		y_0	97.30	0.0265	157.25	0.3423	n.a.	n.a.	n.a.	n.a.
0.9	19/23	y_L	102.5	unst.	unst.	unst.	82.03	unst.	unst.	unst.
		y_0	82.03	unst.	unst.	unst.	unst.	unst.	unst.	unst.

4.7.3. Comparison of LSCEM and LDM for Numerical Simulations

When LDM and LSCEM are compared for the numerical simulations, the damped frequencies obtained with LSCEM differ by less than 1.6% with respect the frequencies obtained with LDM, apart from the case with $p_s = 0.3$ MPa. The estimations of damping factors differ by much more—up to 80%. This parameter is more sensitive to variations in the air gap and pressure.

5. Comparison between Experimental and Numerical Results

The first critical speed of the spindle was experimentally measured by analyzing the amplitude of the synchronous whirl at different rotational speeds. It was found to be 72 krpm and 94 krpm at gauge supply pressure $p_s = 0.5$ and 0.7 MPa, respectively. If compared with the theoretical frequencies resulting from the simplified model of Section 4.5, the difference is less than 6%.

The experimental stiffness on the nose at different pressures is lower than the theoretical value calculated near the rotor-centered condition (between 27% and 39% with respect to the theoretical value). Considering that the 30 N external load imposed on the nose involves non-negligible eccentricity ratios on bearings (especially on the front bearing and with low pressure), a non-linear estimation of the stiffness is preferred (see Section 4.4). If compared with the stiffness obtained in this case, the experimental stiffness is more in accordance with the prediction (between 52% and 86% with respect to the theoretical value).

The experimental damped frequencies and the damping factors evaluated on the shaft vertical vibration at $z = z_L$ are compared with the numerical frequencies in Figure 16. The experimental damped natural frequencies are in good accordance with the numerical frequencies, as they fall in the range defined on the basis of the tolerance considered on the front ($h_f = 17.5 \pm 1.5$ μm) and rear ($h_r = 21.5 \pm 1.5$ μm) clearances. Conversely, the computed experimental damping factors are higher than the numerical simulations. The single-DOF methods were found to be less accurate than LSCEM to capture the modal parameters of experimental signals, especially the damping factors.

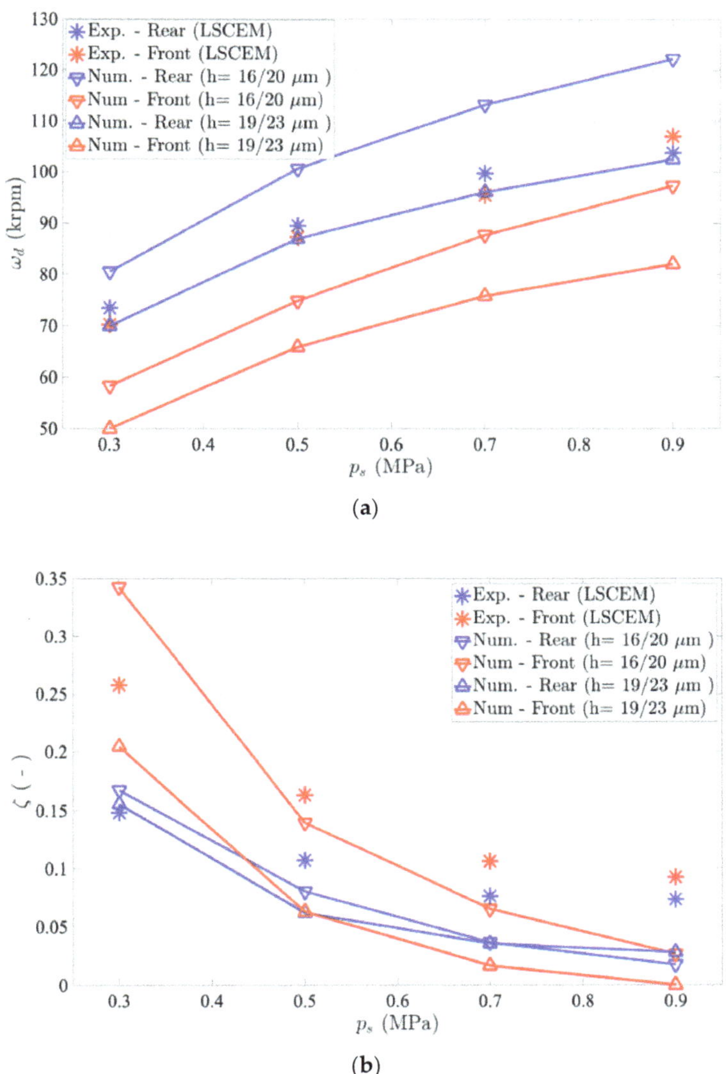

(a)

(b)

Figure 16. Comparison of experimental and numerical damped frequencies (**a**) and damping factors (**b**) on rear plane.

6. Conclusions

This paper describes the experimental and numerical investigation of an electro-spindle supported by aerostatic bearings. The numerical model was validated against literature results in the case of plain aerodynamic bearings and against static and dynamic experimental tests carried out on the electro-spindle at null rotational speed. Both single-DOF methods and a multi-DOF method were proposed to estimate the damped natural frequencies and the damping factors. The accuracy of the model results with respect to the experimental data was discussed. In particular, the damping factor evaluated with LSCEM was in good agreement with the experimental data, while the single-DOF methods overestimated it. Future investigations will regard the dynamic behavior of the spindle at different rotational speeds in order to study its unbalance response and stability.

Author Contributions: Conceptualization, F.C., L.L., A.T., T.R. and V.V.; methodology, F.C. and L.L.; software, F.C. and L.L.; writing—review and editing, F.C., L.L. and A.T.; supervision, V.V.; project administration, T.R.; funding acquisition, T.R. All authors have read and agreed to the published version of the manuscript.

Funding: This research was funded by Carbomech Company.

Institutional Review Board Statement: Not applicable.

Informed Consent Statement: Not applicable.

Conflicts of Interest: The authors declare no conflict of interest.

Nomenclature

$A_{r:\,ij}$	r^{th} modal costant
C	Bearing radial clearance
D	Journal bearing diameter
dt	Time step
d_s	Supply holes diameter
e	Static rotor unbalance
F_c	Bearing reaction force
F_{ext}	External force applied on rotor
f_s	Sampling frequency
g_{in}	Input flow per unit surface through the supply orifices
$h_{i,j}$	Impulse-response function (IRF)
h_f	Radial-film thickness for the front journal bearing
h_r	Radial-film thickness for the rear journal bearing
I	Transversal inertia moment of rotor
I_p	Polar inertia moment of rotor
k	Radial stiffness of journal bearings
k_f	Radial stiffness of the front journal bearing
k_{nose}	Radial stiffness of the spindle evaluated on the nose
k_r	Radial stiffness of the rear journal bearing
k_{ϑ_f}	Tilting stiffness of front journal bearing with respect to its center
k_{ϑ_r}	Tilting stiffness of rear journal bearing with respect to its center
L	Journal bearing length
$l_f,\ l_r$	Axial distance between the rotor center of mass and the centers of the front and rear journal bearings
M_c	Reaction moment in bearings
m_r	Mass of rotor
N	Order of the least squared complex exponential (LSCE) fitting
p	Absolute pressure
p_a	Ambient pressure
p_s	Supply absolute pressure
r	Journal bearing radius
s_r	r^{th} pole of the system
x	Generic position measured along the x-axis
x_G	Center-of-mass position measured along the x-axis
x_{nose}	Spindle nose position measured along the x-axis
x_1	Position measured along the x-axis in the front plane
x_2	Position measured along the x-axis in the rear plane
y	Generic position measured along the y-axis
y_G	Center-of-mass position measured along the y-axis
y_{nose}	Spindle nose position measured along the y-axis
y_{rec}	Signal reconstructed by means of the identified modal parameters
y_1	Position measured along the y-axis in the front plane
y_2	Position measured along the y-axis in the rear plane
z	Generic position measured along the z-axis
z_f	Front-bearing center axial coordinate
z_G	Rotor center-of-mass axial coordinate

z_L	Axial coordinate of the rear end of the rotor
z_{nose}	Spindle nose position measured along the z-axis
z_r	Front-bearing center axial coordinate
z_0	Axial coordinate of the front end of the rotor
z_1	Axial position of the front measuring plane
z_2	Axial position of the rear measuring plane
$\overline{W}$	Dimensionless load capacity
t	Time
ΔT	Time inteval used for LDM
ϵ	Eccentricity ratio
ϕ	Phase angle
φ	Angle identifing dynamic unbalance of the rotor
Φ	Attitude angle
Λ	Bearing number
μ	Air viscosity
γ	Dynamic unbalance of the rotor
$\vartheta_x,\ \vartheta_y$	Rotations around x and y axes
τ	Shear stress
ζ	Damping factor
ω	Angular speed
ω_{con}	Conical mode-shape frequency of the spindle
ω_{cyl}	Cylindrical mode-shape frequency of the spindle
ω_d	Damped natural frequency
ω_n	Undamped natural frequency

Appendix A

Tables A1–A4 report both the natural frequencies resulting from the FFT analysis, both the frequencies and the damping factor of the signal resulting from the logarithmic decrement method.

Table A1. Damped frequencies and damping factors resulting from FFT analysis and from LDM; $p_s = 0.3$ MPa.

h_f/h_r (μm)	T (ms)	Initial Condition	Signal	ω_d LDM (krpm)	ζ LDM (krpm)	$\omega_{d,\,1/2}$ (krpm)
16/20	5	1	y_0	55.556	0.345	49.80 82.03
			y_L	78.125	0.219	
			y_G	55.046	0.361	
			θ_x	56.250	0.325	
	5	2	y_0	55.556	0.346	49.80 82.03
			y_L	78.125	0.220	
			y_G	55.046	0.361	
			θ_x	56.426	0.326	
19/23	5	1	y_0	49.180	0.171	48.40 68.80
			y_L	69.971	0.151	
			y_G	49.180	0.174	
			θ_x	49.315	0.167	
	5	2	y_0	49.180	0.171	48.00 71.60
			y_L	70.175	0.151	
			y_G	49.180	0.174	
			θ_x	49.315	0.167	

Table A2. Damped frequencies and damping factors resulting from FFT analysis and from LDM; $p_s = 0.5$ MPa.

h_f/h_r (μm)	T (ms)	Initial Condition	Signal	ω_d LDM (krpm)	ζ LDM (krpm)	$\omega_{d,1/2}$ (krpm)
16/20	5	1	y_0	74.534	0.119	73.60 100.8
			y_L	99.448	0.100	
			y_G	73.846	0.115	
			θ_x	75.710	0.125	
	5	2	y_0	74.257	0.143	73.20 101.1
			y_L	99.291	0.078	
			y_G	73.892	0.142	
			θ_x	75.567	0.145	
19/23	5	1	y_0	65.934	0.063	66.00 86.60
			y_L	86.331	0.027	
			y_G	65.934	0.065	
			θ_x	65.753	0.059	
	5	2	y_0	65.934	0.063	66.00 86.60
			y_L	86.331	0.027	
			y_G	65.934	0.065	
			θ_x	65.753	0.059	
	10	1	y_0	65.934	0.059	66.00 86.60
			y_L	80.488	0.073	
			y_G	65.854	0.059	
			θ_x	65.854	0.060	
	10	2	y_0	65.934	0.059	66.00 86.60
			y_L	80.685	0.071	
			y_G	65.854	0.059	
			θ_x	65.854	0.060	

Table A3. Damped frequencies and damping factors resulting from FFT analysis and from LDM; $p_s = 0.7$ MPa.

h_f/h_r (μm)	T (ms)	Initial Condition	Signal	ω_d LDM (krpm)	ζ LDM (krpm)	$\omega_{d,1/2}$ (krpm)
16/20	5	1	y_0	87.72	0.053	87.6 113.4
			y_L	112.9	0.052	
			y_G	87.21	0.052	
			θ_x	88.50	0.057	
	5	2	y_0	87.60	0.047	87.6 113.4
			y_L	113.2	0.042	
			y_G	87.27	0.046	
			θ_x	88.13	0.049	

Table A3. *Cont.*

h_f/h_r (µm)	T (ms)	Initial Condition	Signal	ω_d LDM (krpm)	ζ LDM (krpm)	$\omega_{d,\,1/2}$ (krpm)
			y_0	87.464	0.064	
	10	1	y_L	113.6	0.041	87.6
			y_G	87.71	0.034	113.4
			θ_x	114.13	0.053	
			y_0	86.44	0.049	
	10	2	y_L	113.3	0.047	87.6
			y_G	95.00	0.038	113.4
			θ_x	113.5	0.035	
			y_0	75.71	0.019	
	5	1	y_L	95.24	0.010	75.60
			y_G	75.71	0.019	96.00
			θ_x	75.95	0.019	
			y_0	75.76	0.019	
	5	2	y_L	84.41	0.026	75.60
			y_G	75.76	0.019	96.00
19/23			θ_x	75.85	0.019	
			y_0	75.95	0.021	
	10	1	y_L	95.74	0.031	75.60
			y_G	75.71	0.014	96.00
			θ_x	76.19	0.028	
			y_0	75.76	0.020	
	10	2	y_L	96.33	0.031	75.60
			y_G	75.85	0.015	96.00
			θ_x	75.76	0.028	

Table A4. Damped frequencies and damping factors resulting from FFT analysis and from LDM; $p_s = 0.9$ MPa.

h_f/h_r (µm)	T (ms)	Initial Condition	Signal	ω_d LDM (krpm)	ζ LDM (krpm)	$\omega_{d,\,1/2}$ (krpm)
			y_0	97.297	0.024	
	5	1	y_L	121.390	0.036	97.40
			y_G	97.297	0.024	122.0
			θ_x	97.826	0.025	
16/20			y_0	97.297	0.031	
	5	2	y_L	122.245	0.023	97.40
			y_G	97.035	0.013	122.0
			θ_x	124.030	0.040	

Table A4. *Cont.*

h_f/h_r (µm)	T (ms)	Initial Condition	Signal	ω_d LDM (krpm)	ζ LDM (krpm)	$\omega_{d,\,1/2}$ (krpm)
	10	1	y_0	97.335	0.025	
			y_L	122.160	0.025	97.40
			y_G	97.222	0.024	122.0
			θ_x	97.561	0.025	
	10	2	y_0	97.110	0.028	
			y_L	122.450	0.019	97.40
			y_G	97.335	0.019	122.0
			θ_x	123.010	0.024	
	5	1	y_0	82.645	−0.001	
			y_L	87.209	−0.019	82.80
			y_G	82.645	−0.001	102.6
			θ_x	82.873	−0.002	
	5	2	y_0	83.102	0.001	
			y_L	102.560	0.022	82.80
			y_G	82.418	−0.005	102.6
19/23			θ_x	83.565	0.009	
	10	1	y_0	82.759	−0.002	
			y_L	85.511	−0.007	82.80
			y_G	82.759	0.002	102.6
			θ_x	82.854	−0.002	
	10	2	y_0	82.854	0.001	
			y_L	103.330	0.035	82.80
			y_G	82.664	−0.004	102.6
			θ_x	82.949	0.004	

References

1. Bryan, J. International status of thermal error research. *CIRP Ann.* **1990**, *39*, 645–656. [CrossRef]
2. Nishikawa, F.; Yoshimoto, S.; Somaya, K. Ultrahigh-speed micro-milling end mill with shank directly supported by aerostatic bearings. *J. Adv. Mech. Des. Syst. Manuf.* **2012**, *6*, 979–988. [CrossRef]
3. Gill, D.D.; Ziegert, J.C.; Jokiel, B.; Pathak, J.P.; Payne, S.W. *Next Generation Spindles for Micromilling*; Department of Energy: Albuquerque, MN, USA, 2004.
4. Knapp, B.; Arenson, D. Dynamic characterization of a micro-machining spindle. In Proceedings of the 10th International Congress on Membrane and Membrane Process ICOMM, Suzhou, China, 20–25 July 2014.
5. Bediz, B.; Gozen, B.A.; Korkmaz, E.; Ozdoganlar, O.B. Dynamics of ultra-high-speed (UHS) spindles used for micromachining. *Int. J. Mach. Tools Manuf.* **2014**, *87*, 27–38. [CrossRef]
6. Jia, C.; Pang, H.; Ma, W.; Qiu, M. Dynamic stability prediction of spherical spiral groove hybrid gas bearings rotor system. *J. Tribol.* **2017**, *139*, 021701. [CrossRef]
7. Al-Bender, F. *Air Bearings: Theory, Design and Applications*; John Wiley & Sons: Hoboken, NJ, USA, 2021.
8. Rentzepis, G.M.; Sternlicht, B. On the Stability of Rotors in Cylindrical Journal Bearings. *J. Basic Eng.* **1962**, *84*, 521–531. [CrossRef]
9. Sternlicht, B.; Poritsky, H.; Arwas, E. Dynamic stability aspects of cylindrical journal bearings using compressible and incompressible fluids. In Proceedings of the First International Symposium on Gas Lubricated Bearings, Office of Naval Research, Washington, DC, USA, 26–28 October 1959; pp. 119–160.
10. Ausman, J.S. Linearized ph stability theory for translatory half-speed whirl of long, self-acting gas-lubricated journal bearings. *J. Basic Eng.* **1963**, *85*, 611–618. [CrossRef]

11. Ng, C.-W. Linearized PH stability theory for finite length, self-acting gas-lubricated, plain journal bearings. *J. Basic Eng.* **1965**, *87*, 559–567. [CrossRef]
12. Zhang, S.J.; To, S. The effects of spindle vibration on surface generation in ultra-precision raster milling. *Int. J. Mach. Tools Manuf.* **2013**, *71*, 52–56. [CrossRef]
13. Zhang, G.; Ehmann, K.F. Dynamic design methodology of high speed micro-spindles for micro/meso-scale machine tools. *Int. J. Adv. Manuf. Technol.* **2015**, *76*, 229–246. [CrossRef]
14. Müller, C.; Greco, S.; Kirsch, B.; Aurich, J.C. A finite element analysis of air bearings applied in compact air bearing spindles. *Procedia CIRP* **2017**, *58*, 607–612. [CrossRef]
15. Liu, W.; Feng, K.; Huo, Y.; Guo, Z. Measurements of the rotordynamic response of a rotor supported on porous type gas bearing. *J. Eng. Gas Turbines Power* **2018**, *140*, 102501. [CrossRef]
16. Hassini, M.A.; Arghir, M. A simplified and consistent nonlinear transient analysis method for gas bearing: Extension to flexible rotors. *J. Eng. Gas Turbines Power* **2015**, *137*, 092502. [CrossRef]
17. Belforte, G.; Raparelli, T.; Viktorov, V. Modeling and identification of gas journal bearings: Self-acting gas bearing results. *J. Tribol.* **2002**, *124*, 716–724. [CrossRef]
18. Viktorov, V.; Belforte, G.; Raparelli, T. Modeling and Identification of Gas Journal Bearings: Externally Pressurized Gas Bearing Results. *J. Tribol.* **2005**, *127*, 548–556. [CrossRef]
19. Hassini, M.A.; Arghir, M. A New Approach for the stability analysis of rotors supported by gas bearings. *J. Eng. Gas Turbines Power* **2014**, *136*. [CrossRef]
20. Dupont, R. Robust rotor dynamics for high-speed air bearing spindles. *Precis. Eng.* **2015**, *40*, 7–13. [CrossRef]
21. Park, J.-K.; Kim, K.-W. Stability analyses and experiments of spindle system using new type of slot-restricted gas journal bearings. *Tribol. Int.* **2004**, *37*, 451–462. [CrossRef]
22. Czolcznyski, K. *Rotordynamics of Gas-Lubricated Journal Bearing Systems*; Springer: New York, NY, USA, 1999.
23. Colombo, F.; Raparelli, T.; Viktorov, V. Externally pressurized gas bearings: A comparison between two supply holes configurations. *Tribol. Int.* **2009**, *42*, 303–310. [CrossRef]
24. Liu, W.; Bättig, P.; Wagner, P.H.; Schiffmann, J. Nonlinear study on a rigid rotor supported by herringbone grooved gas bearings: Theory and validation. *Mech. Syst. Signal Process.* **2021**, *146*, 106983. [CrossRef]
25. Zachariadis, D.C. Insights Into Discrepancies Between Results of Nonlinear and Linear Vibration Analyses of Rigid Rotors on Journal Bearings. In Proceedings of the Turbo Expo: Power for Land, Sea, and Air, New York, NY, USA, 9–13 June 2008; pp. 1237–1246.
26. Belforte, G.; Colombo, F.; Raparelli, T.; Trivella, A.; Viktorov, V. High-speed electrospindle running on air bearings: Design and experimental verification. *Meccanica* **2008**, *43*, 591–600. [CrossRef]
27. Belforte, G.; Raparelli, T.; Viktorov, V.; Trivella, A.; Colombo, F. An experimental study of high-speed rotor supported by air bearings: Test RIG and first experimental results. *Tribol. Int.* **2006**, *39*, 839–845. [CrossRef]
28. Belforte, G.; Colombo, F.; Raparelli, T.; Trivella, A.; Viktorov, V.; Villavicencio, R. High-speed rotors with elliptic gas journal bearings. Presented at the Proc. of the 5th World Tribology Congress, Torino, Italy, 8–13 September 2013.
29. Den Hartog, J.P. *Mechanical Vibrations*; Courier Corporation: North Chelmsford, MA, USA, 1985.
30. Genta, G. *Vibration Dynamics and Control*; Springer: Berlin/Heidelberg, Germany, 2009.
31. Ewins, D.J. *Modal Testing: Theory, Practice and Application*; John Wiley & Sons: Hoboken, NJ, USA, 2009.
32. Belforte, G.; Raparelli, T.; Viktorov, V.; Trivella, A. Discharge coefficients of orifice-type restrictor for aerostatic bearings. *Tribol. Int.* **2007**, *40*, 512–521. [CrossRef]

MDPI

St. Alban-Anlage 66

4052 Basel

Switzerland

Tel. +41 61 683 77 34

Fax +41 61 302 89 18

www.mdpi.com

Applied Sciences Editorial Office

E-mail: applsci@mdpi.com

www.mdpi.com/journal/applsci